International Scientific and Practical Conference "Sustainable Development in the Post-Pandemic Period" (SDPPP-2021)

AA003549

SHS Web of Conferences Volume 126 (2021)

Tallinn, Estonia
9 - 10 November 2021

Editors:

Olha Pokopenko
Mariia Troian
Marina Jarvis

ISBN: 978-1-7138-4079-4

Printed from e-media with permission by:

Curran Associates, Inc.
57 Morehouse Lane
Red Hook, NY 12571

Some format issues inherent in the e-media version may also appear in this print version.

No changes have been made to the content of these proceedings. There may be changes to pagination and minor adjustments for aesthetics.

Printed with permission by Curran Associates, Inc. (2022)

For additional information, please contact EDP Sciences – Web of Conferences at the address below.

EDP Sciences – Web of Conferences
17, Avenue du Hoggar
Parc d'Activité de Courtabœuf
BP 112
F-91944 Les Ulis Cedex A
France

Phone: +33 (0) 1 69 18 75 75
Fax: +33 (0) 1 69 28 84 91

contact-edps@webofconferences.org

Additional copies of this publication are available from:

Curran Associates, Inc.
57 Morehouse Lane
Red Hook, NY 12571 USA
Phone: 845-758-0400
Fax: 845-758-2633
Email: curran@proceedings.com
Web: www.proceedings.com

TABLE OF CONTENTS

PREFACE

The co-chairs of the International Scientific and Practical Conference "Sustainable Development in the Post-Pandemic Period" (**SDPPP-2021**) undertake the honorable mission on behalf of all conference participants to declare that this event made a significant contribution to the development of modern scientific thought and created new scientific and practical ideas on the path of sustainable development in the post-pandemic period.

The conference brought together leading academic scientists, researchers, Ph.D. students, managers, and practitioners to exchange their experiences and research results on all aspects of Business Transformation and Development during and after the pandemic crises. It also provided the interdisciplinary platform to present and discuss the most recent innovations, trends, concerns, and practical challenges encountered and solutions adopted in the fields of Social Science, Entrepreneurship, Innovation, Business, Management and Education for sustainable development.

The conference was held on November 9-10, 2021, on the online platform of Scientific and Technical Organization *Teadmus* in Tallinn, Estonia. The conference was organized and supported by this organization and the Estonian Entrepreneurship University of Applied Sciences.

More than 200 people from almost 20 countries took part in the conference. This conference proceeding selected the best articles in the following areas:

- Energy policy in the post-pandemic period
- Environmental management under the pandemic influence
- Environmental challenges to the economy in the post-pandemic period
- Revolutionary technical solutions for the development of society
- Digital innovation in the post-pandemic economy
- Changes in business strategies driven by pandemic
- Innovations in monetary and fiscal policy in the post-pandemic period
- Social, managerial and legal aspects of decision making in the post-pandemic economy
- International tourism and leisure industry in the post-pandemic period
- Impact of the pandemic on financial, labour, educational and commodity markets
- Changes in the structure of international trade and services
- Ethics and trust in a healthy environment
- Smart city as a supportive environment for the sustainable company

We thank the invited lecturers, conference participants, the organizing committee and reviewers for the exchange of views and systematization of interdisciplinary knowledge for sustainable development in the post-pandemic period!

On behalf of the SDPPP-2021 Scientific and Organising Committee

Dr. Olha Prokopenko, Prof., Co-Chair, Teadmus OÜ, Estonia; Estonian Entrepreneurship University of Applied Sciences, Estonia;

Dr. Marina Järvis, Ass. Prof., Co-Chair, Estonian Entrepreneurship University of Applied Sciences, Estonia;

Dr. Mariia Troian, Ass. Prof., Secretary, Sumy State Makarenko Pedagogical University, Sumy, Ukraine

Statement of Peer review

In submitting conference proceedings to *Web of Conferences*, the editors of the proceedings certify to the Publisher that

1. They adhere to its Policy on Publishing Integrity in order to safeguard good scientific practice in publishing.
2. All articles have been subjected to peer review administered by the proceedings editors.
3. Reviews have been conducted by expert referees, who have been requested to provide unbiased and constructive comments aimed, whenever possible, at improving the work.
4. Proceedings editors have taken all reasonable steps to ensure the quality of the materials they publish and their decision to accept or reject a paper for publication has been based only on the merits of the work and the relevance to the journal.

Title, date and place of the conference

International Scientific and Practical Conference „Sustainable Development in the Post-Pandemic Period" (SDPPP-2021).
November 9-10, 2021.
Tallinn, Estonia

Proceedings editor(s):

Dr. Olha Prokopenko, Prof.,
Teadmus OÜ, Estonia; Estonian Entrepreneurship University of Applied Sciences, Estonia;
Dr. Mariia Troian, Ass. Prof.,
Sumy State Makarenko Pedagogical University, Ukraine;
Dr. Marina Järvis, Ass. Prof.,
Estonian Entrepreneurship University of Applied Sciences, Estonia

Date and editor's signature

15.11.2021

SHS Web of Conferences **126**, 01001 (2021) https://doi.org/10.1051/shsconf/202112601001
SDPPP-2021

Iron Ore Use as a Potential of Sustainable Development of Industrial Regions of Ukraine: Aspects of Environmental Taxation

Larysa Hromozdova[1*], *Adam* Bodyuk[2], *Uriy* Makhortov[3], *Zoya* Baranik[4], *Nelina* Khamska[5], *Nina* Kukhar[6] and *Carlo* Adami[7, 8]

[1]Kyiv National Economic University named after Vadim Hetman, Department of Regions and Tourism, 03057, Victory Avenue, 54/1, Kyiv, Ukraine
[2]Research institution «Resource», 02149, Larisa Rudenko str., 3B, Kyiv, Ukraine
[3]Kyiv National University of Culture and Arts, Department of International Tourism, 01142, Chigorina str.,20, Kyiv, Ukraine
[4]Kyiv National Economic University named after Vadim Hetman, Department of Economic and Mathematical Modeling and Statistics, 03057, Victory Avenue, 54/1, Kyiv, Ukraine
[5]Vinnytsia State Pedagogical University named after Mykhailo Kotsyubynsky, Department of Pedagogics and Professional Training and Management of educational institutions, 21018, Ostrozkoho str.,32, Vinnytsia, Ukraine
[6]Vinnytsia State Pedagogical University named after Mykhailo Kotsyubynsky, Department of Ukrainian Language, 21018, Ostrozkoho str.,32, Vinnytsia, Ukraine
[7]Taras Shevchenko National University of Kyiv, 01033, str Volodymyrska 64/13, Kyiv, Ukraine
[8] National Bank of Italy, Via Nazionale str., 91, Rome, Italy

Abstract. The article investigates the resources potential of Ukraine, which is not used effectively enough for the needs of sustainable economic development of industrial regions. The article aims to identify ways to increase the eco-tax base in order to improve the budget support for sustainable development of the iron ore subsoil use and synergetic development of industrial regions of Ukraine. The normative-legal base of subsoil users' taxation is investigated. The necessity of ecological taxation is substantiated as the consequence of enterprise activity is actual environmental pollution. Applied innovations in the method of calculating the environmental tax are proposed. They consider the factor of ownership of natural resources and the discount change in their value over time, which will ultimately increase tax revenues to the budget. It is recommended to introduce the author's approach in the methods of calculating the environmental tax in the field of iron ore subsoil use, which synergistically affects the development of industrial regions of Ukraine. It is justified that the COVID - crisis is a heavy burden on local and state budgets. Therefore, it is necessary to further optimize the system of environmental taxation in order to increase environmental contributions to the budget. Keywords: sustainable development, subsoil use, iron ore, deposits, environmental pollution, environmental tax, tax rate, economic geology, industrial regions, COVID – crisis.

* Corresponding author: *gromozdovag@gmail.com*

1 Introduction

In Ukraine, state activity in solving the problems of subsoil use is legalized by regulations, notably the National Program for the Development of Mineral Resources of Ukraine for the period up to 2030 (Law № 3268-VI of 21.04.2011). This state document is valid until 2030, so its provisions are relevant for solving mining problems. In addition, it was amended by the Law of Ukraine № 4731-VI of May 17, 2012.

The following document is the Subsoil Code of Ukraine, which is reflected in the Law of Ukraine of July 27, 1994, № 132/94-BP (latest version № 1423-IX of 28.04.2021). This Law defines the essence of economic activity of users of subsoil resources, their relations with state bodies.

The Subsoil Code of Ukraine (Article 4. Subsoil Ownership) stipulates that "Subsoil is the exclusive property of the Ukrainian people and is provided only for use". All issues and problems with subsoil use must be resolved based on the interests of the Ukrainian people. Rational use of subsoil ensures the well-being of the population, provides jobs for residents of industrial regions of Ukraine. Sustainable development of Ukraine's industrial regions is a potential for economic growth and security of the state. The purpose of environmental taxation is to preserve the health of the nation.

The Cabinet of Ministers of Ukraine has developed documents: "Economic audit of the country", "Vectors of economic development until 2030". These documents became the basis of the "National Economic Strategy until 2030". These government documents analyze the current state of the country's economy, present the general directions of government activities to improve the situation in various industries. In these documents, the Government emphasizes the importance of solving the problems of ferrous metallurgy, ore mining and growth of production and resource potential of the mining and metallurgical complex of the country.

The post-pandemic period requires further research. Now we need innovative sound recommendations for the development of the country's mineral resources complex. This will help increase Ukraine's political prestige and international economic importance. In the context of the post-pandemic budget deficit, solving the fiscal obligations of the country's mining and metallurgical complex is very important.

The purpose of the research is to identify ways to increase the eco-tax base in order to improve the budget support for sustainable development of the iron ore subsoil use and synergetic development of industrial regions of Ukraine.

2 Results

According to statistical indicators, Ukraine ranks first in Europe in terms of iron ore reserves of various types. Therefore, the iron ore economy represents a powerful potential for sustainable development of the industrial regions of Ukraine and the country's economic development as a whole. This potential can ensure financial growth in the country, providing jobs for the population of different territories.

The National Program for the Development of the Mineral Resources Base of Ukraine for up to 2030 states that in Ukraine, iron ores (explored and confirmed reserves) amount to 28 billion tons. Globally, iron ore reserves are distributed among four countries: Russia, Ukraine, Brazil and China. Their total share in the world's proven reserves is 67.2 per cent (139 billion tons). Thus, the natural resource potential of meeting the needs of industrial regions of Ukraine, in particular in the field of metallurgy, is too strong.

Thus, the promising development of ferrous metallurgy in Ukraine is due to too large a resource of iron ore, in particular. These ores are divided into three main types: rich magnetite-hematite, ferrous quartzite, oolite brown iron ore.

Iron quartzites belong to the unoxidized type of Precambrian formations of the Ukrainian Shield. They are concentrated in deposits of so-called poor martite ores. They have an iron content of 15-45%. In the oxidized zone, rich ores of hematite composition have 46-70% iron. Main with iron quartzites deposits of Kryvyi Rih iron ore basin, Kremenchug and Belozersky iron ore districts [4].

According to the content of ore reserves, the following deposits of ferrous quartzites are distinguished: unique resource (over 10 billion tons) and significant resource (over 1 billion tons).

Economic geology (EG) and resource geology (RGL) is engaged in research of problems of information and cost reflection of a resource geological condition.

3 Analysis of recent research and publications

Problems of the economic geology of iron ores and their deposits, in particular iron quartzites, are most widely covered in scientific works by the following authors: N.V. Vinnik, M.N. Korzhnev, M.M. Kurilo, M.H. Palilo, O.Petrusenko, O.V. Plotnikov, S.V. Radovanov, G.I. Rudkom and other scientists of geological sciences. Their research focuses on economic assessments of ore deposits, performance indicators of their exploitation, in particular iron. Thus, M.M. Kurilo, O.V. Plotnikov, S.V. Radovanov, G.I. Rudko conducted research and reflected the methodological foundations of geological and economic assessments of ferrous quartzites and their deposits, analyzed the compliance of the mineral resource base to the requirements of existing mining iron ore enterprises [4-6].

Iron ore is a dominant resource that can ensure the sustainable development of the industrial regions of Ukraine.

Officially, the State Program of Ukraine stipulates that in order to meet the demand for naturally rich iron ores in 2021-2030, the domestic industry needs an increase in their production over these years by 450 million tons. They are receiving 2600 million tons by conducting a searching assessment of Hrushkiv, Sekretarsky and Novoselytsky manifestations, exploration of Pishchansky, Savransky, Baibuzovsky and Moldovan deposits in the Middle Pobuzhye and others. The needs for financing the implementation of this program, in particular iron ore facilities, have been identified.

The dynamics of indicators of production of ores and concentrates of agglomerated iron are given in the table. 1.

Table 1. Production of ores and agglomerated iron concentrates

Years	2015	2016	2017	2018	2019	Average figure
Production, million tons	55,2	56,7	50,8	53,1	51,7	53,5
Chain indices	1,0	1,03	0,90	1,04	0,97	-

Source [7, p. 254]

They are listed in table. 1 data reflect the production of ores and iron agglomerated concentrates in the pre-pandemic period.

Exports of extracted ores and concentrates in 2018 reached 36.9 million tons, amounting to 2869.0 million dollars. USA; in 2019 - 39.9 million tons, amounting to 3397.8 million dollars. USA [8, p. 51].

The cost of resources and reserves of minerals in individual deposits or subsoil areas is calculated according to the Methodology for determining the value of reserves and resources of minerals of the deposit or subsoil area. This Methodology is approved by the Resolution of the Cabinet of Ministers of Ukraine of August 25, 2004, № 1117.

According to the substantiations of Gosgeonadr of Ukraine, the cost of resources and reserves of minerals in individual deposits or subsoil areas is recommended to be calculated by the formula in the following wording:

$$Bp = \sum [(Dt - Bt) - Pt] + A)/(1+E)^t - \sum Kt/(1+E)^t \qquad (1)$$

where *Bp* - the cost of resources on the date of assessment;

E – discount rate;

Dt – the amount of annual income from the sale of marketable products, which is accounted for as an object of rent, in the *t*-th year;

Bt – the number of operating costs in the *t*-year, which includes the amount of depreciation of fixed assets and costs associated with economic activities for the extraction of minerals (called the production cost of marketable products of the entity);

Pt – the amount of taxes and payments in the t-year, which are not included in operating costs;

At – the amount of depreciation of fixed assets and costs associated with economic activities for the extraction of minerals in the *t*-th year;

Kt – capital investments in industrial construction in the *t*-th year, including the acquisition of geological information;

T – the period of use of the deposit or subsoil area for geological study and/or extraction of minerals before decommissioning of the deposit;

Σ – sum sign [9].

In our opinion, it is not objective. For example, the discount rate used to determine the value of resources is taken equal to the discount rate of the National Bank of Ukraine. The formula indicates the *t*-th year. However, the discount rate of the National Bank of Ukraine changes several times, even from any specific monthly dates of one year.

This formula, as well as other authors' improvements, does not take into account the factor of national ownership of mineral deposits.

In addition, it is logical to use the proposal of V.V. Matyukha to take into account the discount, which changes the cost of the resource over time. He recommended the inclusion in the above formula of a correction factor (P), which estimates the relationship between the original cost of the resource and the value, together with interest discounts, at the end of the period. Thus, the author proposes to take into account the value of the resource (its monetary value) over time [10, p. 116].

Taking into account the proposal of V.V. Matyukha, the denominator of this formula will have the following form [10, p. 116]:

$$(1 + E * P)^t \qquad (2)$$

For many years, the Tax Code of Ukraine used a single rate of 8% of the subsoil use fee for iron ore mining. But the quality of these ores, mining conditions and other factors are different and depend on their deposits. Recently, variable rent rates for iron ore mining have been introduced, the value of which already depends on market factors.

If the average value of iron ore, according to the IODEX index 58% FE CFR China, which is officially determined by the world news agency Platts, for the tax (reporting) period is 70 and more US dollars, then the rate is 12.00%.

If the average value of iron ore, according to the IODEX index 58% FE CFR China, which is officially determined by the world news agency Platts, for the tax (reporting) period is less than 70 US dollars, then the rate is 11.00%.

In our opinion, these two orientations are not enough for the objectivity of the fiscal obligations of mining companies, as it is essential to take into account the average cost of iron ore.

Thus, in assessing the financial results of subsoil use, it is necessary to introduce further market orientations of the economic geology of iron ore.

However, it should be noted that the quality of iron ore, the conditions of their extraction, profitability, environmental consequences - differ by territory. Therefore, the rate of environmental tax for pollution should be set depending on the fiscal geological policy.

The dynamics of rates for calculating the environmental tax for the presence of hazardous waste, according to the Tax Code of Ukraine, are given in the table. 2.

These data show an increase in such a short period of environmental tax by 69%.

Table 2. Dynamics of rates for calculating the environmental tax

Danger Class	Waste hazard level	Tax rate, UAH per 1 ton in 2014	Tax rate, UAH For 1 ton in 2021	Rates 2021 to 2014, %
I	Extremely dangerous	822,52	1405,65	170,9
II	Highly dangerous	29,96	51,20	170,9
III	Moderately dangerous	7,52	12,84	170,7
IV	A little dangerous	2,93	5,00	170,6
V	Low-hazardous non-toxic waste from the mining industry	0,29	0,49	169,0

Source: [1, 7]

The scale of payment of the environmental tax by economic entities is evidenced by the following data. Private Joint-Stock Enterprise "Northern Mining and Processing Plant" (PJSC "PIVNGZK", Kryvyi Rih), engaged in the extraction of iron ore (07.10 Extraction of iron ore), in 2020 had an income of 24,063 bln UAH, paid environmental taxes in the amount of 151,674 mln UAH (its authorized capital is 579,707 bln UAH), which is 26% of the authorized capital [12; 15].

Public Joint Stock Company "Zaporizhzhya Metallurgical Plant" (PJSC "ZAPORIZHSTAL", Zaporizhia), engaged in the production of cast iron, steel and ferroalloys (24.10 Production of pig iron, steel and ferroalloys), had an income of 45,630 bln UAH paid eco 146,087 mln UAH (its authorized capital is 660,920 mln UAH), which is 22% of the authorized capital [13; 15].

Public Joint Stock Company "ARSELORMITTAL KRYVY RIG" (PJSC "ARSELORMITTAL KRYVY RIG", Kryvyi Rih) carries out the following activities: 24.10 Production of pig iron, steel and ferroalloys; 07.10 Extraction of iron ores; 19.10 Production of coke and coke products, etc. Its authorized capital is 3,859 bln UAH. It paid environmental taxes in the amount of 316,771 mln UAH, which is 8.2% of the authorized capital [14; 15].

In addition to the applied aspects of the economics of nature management, it is proposed to introduce new concepts in the theory of the economics of nature management, subsoil use and in the theory of their taxation. In particular, supplement the definition of "environmental tax" with the definition of "environmental tax" and distinguish them by content and function. In the case of total payment, an environmental tax should be used, in the case of a penalty - an environmental penalty. The introduction of these definitions will ensure the harmonization of terminological interpretations in the economics of nature management, subsoil use, tax theory and tax legislation in general.

4 Conclusions and suggestions

Ukraine is one of the richest countries in the world in terms of ore and resource potential. However, this potential is not used effectively enough for the needs of sustainable economic development of the industrial regions of Ukraine.

The regulatory framework for estimating the cost of subsoil uses needs to be improved, in particular, to take into account a range of natural and economic factors.

Theoretical provisions of ecological geology, fiscal geology, economic geology of iron ore, as well as other minerals must be combined in economic assessments of mining regions and enterprises. At the same time, it is necessary to different indicators and fiscal rates of taxation of enterprises depending on the factor of natural features of mineral deposits extracted by these enterprises. It is necessary to introduce along with the "environmental tax" payment called "environmental penalty" and distinguish them by content and function.

The current mechanism of environmental taxation of subsoil users, especially the calculation of environmental tax, needs to be improved.

This is relevant for the pandemic period. Subsoil users should be considered as economic entities whose activities result in environmental pollution. It is necessary to introduce innovations in the method of calculating the environmental tax, which will take into account the factor of national ownership of natural resources. It is recommended to implement the proposed author's approach in the methods of calculating the environmental tax.

Of course, the COVID crisis is a heavy burden on local and state budgets. Therefore, it is necessary to optimize the system of environmental taxation further and increase environmental contributions to the budget of subsoil users, in particular, iron ore deposits. This is a prerequisite for sustainable development and the subsoil use system, and the industrial regions of Ukraine as a whole.

When calculating the environmental payment, it is necessary to take into account the factor of privatization of water resources. The eco-tax of mining companies that have caused water pollution is becoming a source of recovery of these resources in the future on the principles of sustainable development.

In the conditions of decentralization, territorial communities should intensify the actual implementation of environmental measures and form eco-funds for environmental contributions of geological and mining entities in order to ensure sustainable development.

The funds generated should be used optimally to finance measures to reduce the level of locally manufactured load, to restore the resources used by businesses and the public, in particular, in the context of COVID - a protracted crisis.

When calculating the environmental payment, it is necessary to take into account the factor of privatization of water resources. The eco-tax of mining companies that have caused pollution of private water resources is becoming a source of recovery of these resources in the future on the principles of sustainable development.

In the conditions of decentralization, territorial communities should intensify the actual implementation of environmental measures and form eco-funds for environmental contributions of geological and mining entities, in order to ensure sustainable development.

The funds should be used optimally to finance measures to reduce the level of local manufactured load, to restore the resources used by businesses and the public, in particular, in the context of COVID - a protracted crisis.

References

1. Law № 3268-VI, National Program for the Development of Mineral Resources of Ukraine for the period up to 2030. URL: https://zakon.rada.gov.ua/laws/show/3268-17#Text
2. Subsoil Code of Ukraine: Law of Ukraine of July 27, 1994 № 132/94-VR, as amended. and add., the last № 1423-IX from 28.04.2021. URL: http://sfs.gov.ua/zakonodavstvo/podatkove-zakonodavstvo/zakoni-ukraini/arhiv-zakoniv-ukraini/zakoni-ukraini-za-1994-rik/30627.html

3. Resolution of the Cabinet of Ministers of March 3, 2021 № 179 On approval of the National Economic Strategy for the period up to 2030: URL https://www.kmu.gov.ua/npas/pro-zatverdzhennya-nacionalnoyi-eko-a179.
4. M.M. Korzhnev, V.A. Mikhailov, B.C. Mishchenko. *Fundamentals of economic geology*, KL, 223 (2006)
5. O.B. Bobrov, M.D. Krasnozhon, B.I. Malyuk. *Subsoil use in Europe and America*: Reference edition. KG, 197 (2003)
6. M.M. Kurilo, S.V. Radovanov, G.I. Rudko. *Geological and economic assessment of mineral deposits*. ADEFU, 384 (2011)
7. Statistical Yearbook of Ukraine for 2019. URL: http://ukrstat.gov.ua/druk/publicat/kat_u/2020/zb/11/zb_yearbook_2019.pdf
8. Foreign trade of Ukraine: statistical yearbook 2020. URL: http://www.ukrstat.gov.ua/druk/publicat/kat_u/2020/zb/06/Zov_torg.pdf.
9. On amendments to the Methodology for determining the value of reserves and resources of minerals of the deposit or subsoil area provided for use. URL: ww.drs.gov.ua/wp-content/uploads/2018/07/10137_19-18.pdf.
10. V.V. Matyukha. *Methodological approaches to improving the method of discounting cash flows,* in Proceedings of the international scientific conference Scientific principles of geological and economic assessment of mineral resources of Ukraine and the world. NC, 12-13 April 2011, Kyiv, Ukraine (2011).
11. Tax Code of Ukraine. URL: https://zakon.rada.gov.ua/laws/show/2755-17.
12. Database of Ukrainian Enterprises. URL: https://youcontrol.com.ua/catalog/company_details/00191023/.
13. Open Data Bot. URL: https: //opendatabot.ua/c/00191230.
14. Company details in the National Database. URL: https://youcontrol.com.ua/catalog/company_details/24432974/.
15. Information on payment to the Consolidated Budget of Ukraine by payers - subjects of natural monopolies and payers of rent for subsoil use. URL: https://www.saveecobot.com/analytics/ecotaxes

Innovation Policy Coherence for Sustainable Development

Vitaliy Omelyanenko[1,2], *Olha Prokopenko*[1], and Olena Omelyanenko[2] *

[1]Teadmus OÜ, Estonia
[2]Makarenko Sumy State Pedagogical University, Ukraine
[3]Institute of Industrial Economics of NAS of Ukraine, Ukraine

Abstract. The research deals with the conceptual framework of innovation policy coherence for sustainable development. Authors prove that for the development of the national system, «sustainable development – security» should be considered innovation policy as a set of management tools for targeted implementation of innovations that will promote sustainable development. As a result of understanding the importance of SDGs, a scientific and practical approach to policy coherence for sustainable development (PCSD) is considered. Given the importance of innovation development and the national characteristics of the relationship between it and the SDGs, it is advisable to consider the methodology of coherence of innovation policy for sustainable development (innovation policy coherence for sustainable development, IPCSD). The context of innovation policy coherence for sustainable development based on ideas of theorists of economic science, innovation and strategic management is presented. It is proposed to implement the conceptual provisions of innovation policy coherence for sustainable development on the basis of adapted principles of sustainable development and principles of the relationship «sustainable development – security». Keywords: innovation policy, sustainable development, coherence, strategic management, security.

1 Introduction

The security problem is the main one on the list of priorities of the country's strategic development. It covers the complexity of transformation processes at the beginning of the XXI century and is a condition and goal of development policy, as it requires the concentration of all forces and resources on solving problems. Their diversion to neutralize threats complicates, and sometimes makes impossible, the successful implementation of development strategies. The existence of two primary and closely interrelated functions of the state – promoting socio-economic development and national security – raises the question of them in a closer union in a single functional system «development – security» in the transition to sustainable development as a basis for security.

To conceptualize the link between sustainable development and security, particularly

* Corresponding author: elnikkrasnaya@gmail.com

within the sustainable development goals (SDGs), and to develop appropriate policies, it is necessary to find the most effective management tools, one of which, research has shown, is innovation policy.

Therefore, the purpose of this study is to determine the conceptual framework of innovation policy coherence for sustainable development.

2 Innovation component of sustainable development

The value of the innovation component is illustrated by the results of dispersion decomposition and identification of differences in how each SDG contributes to explaining the variance in the welfare of countries (Figure 1), according to which SDG 9 has the most considerable contribution (13%) among other SDGs. The importance of SDG 9 can also be illustrated by the relationship between SDGs [1].

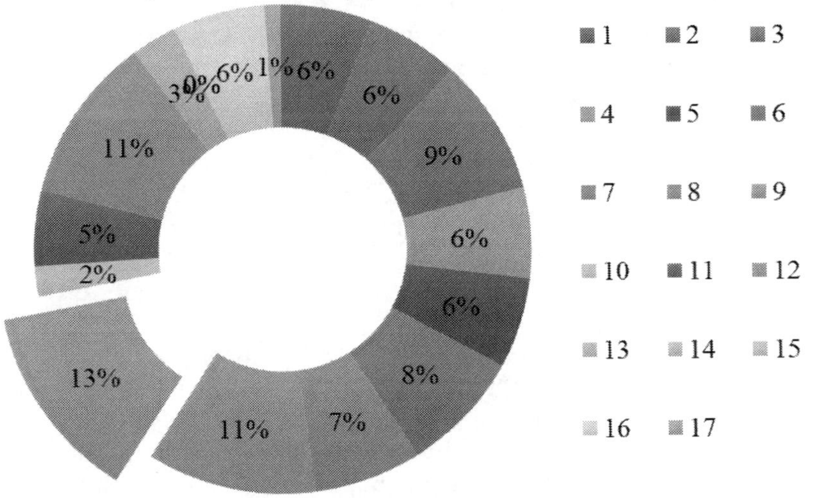

Fig. 1. The relative importance of SDGs to explain differences in the welfare of countries
Source: World Happiness Report 2020 [2]

SDGs depend on the development and dissemination of innovative technologies. From this point of view, innovation and creativity are not in themselves goals, but means and tools for creating solutions to development problems and, being at the centre of the system, affect many SDGs.

Based on the above, for the development of the national system, «sustainable development – security» should be considered innovation policy as a set of management tools for targeted implementation of innovations that will promote sustainable development. The need to analyze the innovative dimension of the national system «sustainable development – security» is because the narrowing of national security to the military-political aspect concerning national interests in security and defence is a wrong approach in current conditions. In this context, innovation security can be a component that connects research in the interests of both national security and its high-quality socio-economic development.

In the context of achieving the SDGs should also take into account the results of the analysis of innovation policy, considered in the study [3]:

1) the policy of innovation development should be based on a broader definition of innovation, recognizing the importance of organizational and social innovations and innovations in public administration, but at present, it remains focused mainly on technology; the same applies to the existing innovation model, which mainly reflects the

linear way of innovation, exaggerating the importance of research compared to the daily practice of firms;

2) policy should recognize the importance of involving stakeholders in the innovation process and seek to identify a wide range of actors and build effective partnerships around essential topics;

3) internal organizational dynamics are poorly considered in innovation policy, partly due to its technological orientation and linear model of innovation;

4) politics is often a top-down approach, and the interconnected nature of players and processes at the micro, meso and macro levels is poorly defined.

As a result of the integrated nature of SDGs, new innovative approaches are discussed in a study [4] based on the following broad strategies:

– mission-oriented innovations include the implementation of network research programs at the national and international levels, as well as the creation of unique organizational structures that can guide innovation to achieve specific SDGs;

– pro-poor and inclusive innovations: expanding the benefits of innovation and building on the ideas of innovation for the bottom of the pyramid of needs through new concepts, inexpensive labour and large-scale production;

– grass-roots innovation: expanding the range of subjects of the innovation process to attract grass-roots innovation movements; this approach focuses on the practice of innovation in both technology and service delivery;

– social innovation: the transition from technological to social innovation, focusing on organizational innovation and new social practices aimed at improving welfare (for example, in business models, industrial practices and finance and the provision of public services);

– open and joint innovations with digital support: promotion of open digital cooperation; such innovative approaches absorb and recombine many sources and forms of knowledge, primarily through the digital support of open collaboration.

As a result of understanding the importance of SDGs, a scientific and practical approach to policy coherence for sustainable development (PCSD) can be considered. OECD experts define PCSD as «an approach and tool for the systematic integration of economic, social and environmental aspects of sustainable development at all stages of domestic and international policy-making» [5]. The main goals of PCSD are as follows:

– promoting synergies in all areas of economic, social and environmental policy;

– identification of compromises and harmonization of domestic policy goals with international goals;

– elimination of the negative consequences of domestic policy for the SDGs.

Given the importance of innovation development, as well as the national characteristics of the relationship between it and the SDGs, it is advisable to consider the methodology of coherence of innovation policy for sustainable development (innovation policy coherence for sustainable development, IPCSD).

Accordingly, the security of the innovative component of sustainable development can be defined as the level and dynamics of innovative development that protect national interests from threats arising from domestic and global trends of socio-economic processes and targeted actions of foreign agents, based on competitive results of scientific, technical and innovative activities embodied in products (services).

3 Innovation policy coherence for sustainable development context

We propose to consider the context of innovation policy coherence for sustainable development (IPCSD) on the basis of such ideas of theorists of economic science, innovation and strategic management.

1. Schumpeter's ideas about competition based on innovation and their analysis as a major factor in economic development. Within this approach, innovations are the basis of competition and a factor in shaping the structure of industry markets. Accordingly, we can consider innovation as a factor in the dynamics of markets of the national economy, which, depending on the effectiveness of coordination management, creates opportunities or threats.

2. Hayek's concept, within which the competitive market is interpreted as a special information environment that identifies, uses and coordinates the use of knowledge of many independent participants.

3. The concept of Freeman's national innovation system. According to this concept, the national innovation system is a set of institutions that individually and comprehensively implement the functions of creating and disseminating new technologies, forming the basis for the formation and implementation by the government of policies that affect the innovation process. In developed countries, effective mechanisms ensure high market efficiency and low transaction costs. Due to such mechanisms, competition in these countries is based on new knowledge and technologies (search for knowledge according to the concept of Hayek), which changes the logic of processes in the economy and innovation system.

4. Understanding of innovation policy within a number of schools of strategic planning, linking the following aspects: reaction to changes in the external environment (reactive process); transition to a new stable position – a configuration with a new set of strategies, structures and principles (proactive process).

5. Kramer's O-ring theory. In the original, the O-ring is a standard part for sealing valves and other moving mechanisms. In 1986, during the launch of the shuttle «Challenger», damage to one of these rings led to the explosion and death of seven astronauts.

Kramer argues that the production of goods is like launching a space shuttle: even creating a simple product in the economy requires the coordinated work of manufacturers, accountants and marketers, etc., and failure at least one stage means that the product will not be successful.

From these positions, the O-ring theory, in particular, explains the differences in GDP per capita between countries, which cannot be explained by differences in the stock of physical or human capital. Due to the fact that in developed countries, it is more profitable to use technologically more complex processes, small differences in the skills of workers through the O-ring mechanism turn into greater differences in productivity between countries. Derived from the theory of the O-ring is the theory of disorganization, developed by Kramer in collaboration with prof. Blanchard. In the production chains of a planned economy, the relationship between firms was often specific: many firms had only one supplier and one buyer.

6. The strategy of the «blue ocean» of K. Chan and R. Moborn, which can be adapted to the needs of the system «sustainable development – security». According to the strategy's authors, the «blue ocean» is a new non-competitive market based on new value. We propose to use the «blue ocean» strategy to build a national system of «sustainable development – security».

4 Principles of innovation policy coherence for sustainable development

It is proposed to implement the conceptual provisions of innovation policy coherence for sustainable development (IPCSD) based on:

– adapted principles of sustainable development;
– principles of the relationship «sustainable development – security».

These principles include the following.

1. The principle of consistency of development strategy with all available strategies (sectoral, territorial, resource). This principle determines the degree of consistency of the step-by-step implementation of the overall innovation strategy with the strategy of safe development.

2. The principle of interaction of the strategy of safe development with the forecasted changes of the external environment (technological dynamics). It should be based on the compliance of the subsystem development strategy with the projected changes in indicators (national and international dimensions) to minimize threats and realize the potential of the external environment.

3. The principle of interaction of the strategy of sustainable, safe development of the state with its innovative capabilities. This principle determines the potential for the formation of resources for innovative development and the compliance of development strategies with resources.

4. The principle of the reality of the development strategy determines the strategic opportunities in the implementation of complex innovation and investment projects in a certain perspective, in the formation of the required amount of resources.

5. The principle of determining the acceptable level of risk (strategic risk corridor of development) associated with the implementation of selected development strategies (set of strategies). Based on this principle, the acceptability of the level of risks for the activity is determined in terms of the possible size of losses and the generation of threats to strategic security.

6. The principle of economic and social efficiency of the development strategy. It shows compliance with the strategic target settings, image, level of control over the activities of functional institutions.

7. The principle of system solutions. SSM in the field of innovation covers activities that combine the interconnected provision of political, economic and military leadership, as well as diplomacy and other political and legal aspects.

8. The principle of technological relationships. Breakthrough innovations can not only significantly change the technical and technological foundations of the industry but also markets, the composition and roles of actors, as well as the foundations of the economy.

9. The principle of the national (local) orientation of innovation development. For development, it is necessary to choose such technological industries that will be related to the existing potential, welfare issues and build a solid foundation for new technologies and areas of development relevant to a particular country (region).

10. The principle of strategic orientation, which provides for the inclusion of the innovation component in programs and development strategies. Strategic priorities set long-term guidelines for the development of innovation entities, as well as guidelines for financing the sector of basic and applied science and support for innovative communications. Strategic management culture is important within this principle.

5 Conclusion

The role of innovation policy in strengthening national security in the context of achieving the SDGs of the national economy is often ignored. Therefore, there is a need to analyze the essence of the innovative component of the relationship between sustainable development and national security. In this context, innovation policy is the component that directs innovation activity to meet the interests of both national security and socio-economic development of the country. Given the current importance of innovation for sustainable development, relevant national characteristics, as well as from an applied point of view, it is important to harmonize innovation policy with the SDGs, which provides support for science and innovation priorities and measurements of sustainable development.

References

1. D. Zelinka, B. Amadei, International Journal of System Dynamics Applications, **8**, 1 (2019)
2. J. F. Helliwell, R. Layard, J. Sachs, Jan-Emmanuel De Neve eds. *World Happiness Report 2020* (New York, Sustainable Development Solutions Network, 2020) URL: https://worldhappiness.report/ed/2020/sustainable-development-and-human-well-being/#fnref21 (2020)
3. C. Makó, M. Illéssy, QuInnE Working Paper, **1** (2015)
4. *New innovation approaches to support the implementation of the sustainable development goals.* United Nations. URL: https://unctad.org/en/PublicationsLibrary/dtlstict2017d4_en.pdf (2017)
5. *Policy Coherence for Sustainable Development 2019. Empowering People and Ensuring Inclusiveness and Equality Highlights.* OECD. URL: https://www.oecd.org/gov/pcsd/policy-coherence-for-sustainable-development-highlights-2019.pdf (2019)
6. V. Omelyanenko, O. Braslavska, N. Biloshkurska, M. Biloshkurskyi, N. Kliasen, O. Omelyanenko, International Journal of Computer Science and Network Security, **21**, 9 (2021)
7. V. Omelyanenko, *National strategic innovation security policy making (theoretical review)* (Tallinn, Teadmus, 2020)
8. V. Omelyanenko, Technology audit and production reserves, **3/5** (41) (2018)

SHS Web of Conferences **126**, 01003 (2021) https://doi.org/10.1051/shsconf/202112601003
SDPPP-2021

Decarbonization of Ukrainian Economy: Economic and Legal Issues

Liubov Zharova[1*], and *Eremeeva* Natalia[2]

[1]Wyższa Szkoła Ekonomiczno – Humanistyczna, Management Department, 43-300 Bielsko-Biała, ul. gen. Wł. Sikorskiego 4-4c, Poland
[2]The Institute of Economic and Legal Research of NAS of Ukraine, Department of economic and legal research of economic security issues, 01032 Kyiv, 60 Shevchenko blvd., Ukraine

Abstract. The research hypothesis is determined by its focus on forming practical proposals for creating a compelling economic and legal mechanism for the decarbonization of the economy. Based on an integrated approach to the regulation of relevant relations. This approach allows us to identify priority areas for the transformation of legal regulation in the context of decarbonization of the economy, which in turn will help lay the legal foundations for the transition to a carbon-free economy. The practical implementation of the study's scientific results will contribute to implementing the tasks of the Renewed Nationally Defined Contribution of Ukraine to the Paris Agreement. It is planned to pay special attention to the legal aspects of modernization of the coal industry in the context of stimulating the reduction of greenhouse gas emissions. Additional attention was paid to the development of theoretical provisions and practical proposals for the harmonization of state policy in the field of bioenergy, its legal framework with the goals of decarbonization of the energy sector, and ensuring climate neutrality. Keywords: decarbonization, bioenergy, carbon-free economy, climate neutrality.

1 Introduction

As the global economy recovers from the immediate economic impact of the Covid crisis, attention is increasingly turning to the long-run effects of the shock on the economy's productive potential and the possibility of not only recovering but renewing in the framework of sustainability. The Beyond Recovery initiatives of the United Nations Development Program exemplify the potential for building a new social contract, uprooting inequalities, rebalancing nature, climate, and the economy, and accelerating and scaling digital disruption and innovation [1].

The global nature of climate change necessitates implementing effective mechanisms that can significantly reduce the risks and negative consequences of climate change at the international and national levels. According to the scientific consensus that becomes even stronger in pandemic times, combating climate change requires full decarbonization of our energy systems and zero net greenhouse gas emissions by around 2070. Due to Paris

* Corresponding author: zharova.l@gmail.com

Agreement [2], countries are negotiating much more modest steps to 2025 or 2030, called Intended Nationally Determined Contributions (INDCs). The long-term goals specified in Articles 2 and 4.1 of the Agreement also establish the time limits. Herefore, it builds on a ratcheting up of aggregate and individual ambition over time. NDCs are submitted every five years to the UNFCCC secretariat. The Paris Agreement provides that successive NDCs will represent a progression compared to the previous NDC and reflect its highest possible ambition to enhance the aim over time. Parties are requested to submit the next round of NDCs (new NDCs or updated NDCs) by 2020 and every five years after that (e.g., by 2020, 2025, 2030), regardless of their respective implementation time frames.

Simultaneously, planned measures and proposed tools for combating climate change were not implemented as efficiently as expected. According to the Climate Action Tracker [3], current global pledges are insufficient (fig. 1).

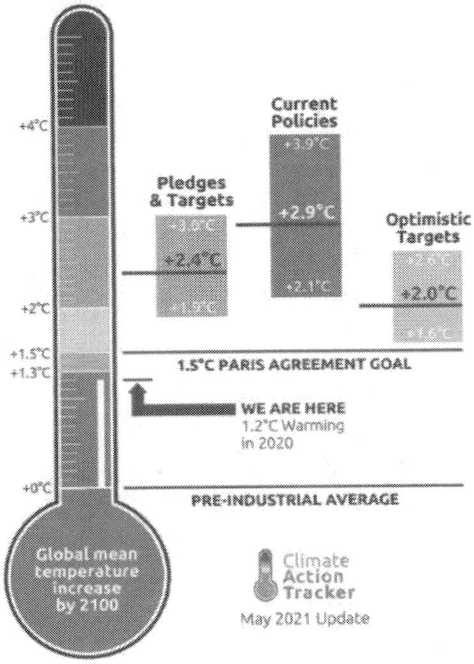

Fig. 1. Climate action announcements and scenarios.

To summarize, decarbonization is part of the global climate change response introduced by several international instruments (the UN Framework Convention on Climate Change, the Kyoto Protocol, the Paris Agreement). It is a tool to mitigate the effects of climate change, is a system of measures to reduce emissions of greenhouse gases to curb the growth of the global average temperature. The commitments made by the member states of the New Global Agreement (limiting the increase of the average temperature of the Earth's atmosphere to 1.5°C above pre-industrial levels (Article 2 of the Paris Agreement [2])) determine the transformation of the world economy towards climate neutrality based on so-called *green energy transition* – replacement of fossil fuel energy systems with renewable electricity and renewable gases (hydrogen, biomethane, synthetic methane); increasing the energy efficiency of national economies. The European Community's response to these challenges has been the European Green Deal [6], a strategy for the transition to a carbon-free economy by 2050. For Ukraine, the realization of the goals of the Paris Agreement is at the same time a direction of fulfilment of obligations in the context of European integration aspirations.

Ukraine submitted its updated NDC on July 31, 2021, significantly strengthening its 2030 target but still falling short of achieving a 1.5°C – compatible ambition level [4]. The NDC further includes the goal of reaching climate neutrality no later than 2060. The submission targets a 65% reduction below 1990 levels by 2030, including LULUCF, slightly more ambitious than the range proposed at the Climate Ambition Summit in December 2020 (58% - 64% below 1990) [5]. The submitted target significantly improves its previous target of at least a 40% reduction below the 1990 level by 2030.

2 Economic and Legal Framework

The Ukrainian Government approved the Renewed Nationally Defined Contribution of Ukraine to the Paris Agreement, reducing greenhouse gas emissions by 65% in 2030 from the 1990 level (fig. 1). Commitments to significantly reduce emissions will be met in the face of long-term negative trends in the economy: the ageing of fixed assets, technological equipment, low levels of technical, technological, and energy efficiency against the background of increasing capital investment deficits. Large-scale structural changes require significant energy consumption. However, the functioning of the country's energy sector is primarily due to the use of fossil fuels - coal, natural gas, and fuel oil. The increase in the share of renewable energy sources is limited by the current state of the energy system, which requires a qualitative transformation, and hence additional capital expenditures. At this stage, Ukraine is not ready to completely abandon fossil fuels (energy, metallurgy, utilities and household). Moreover, possible economic growth, increased production of marketable products (electricity, metal, etc.) will increase the relevant emissions and question the fulfilment of international obligations.

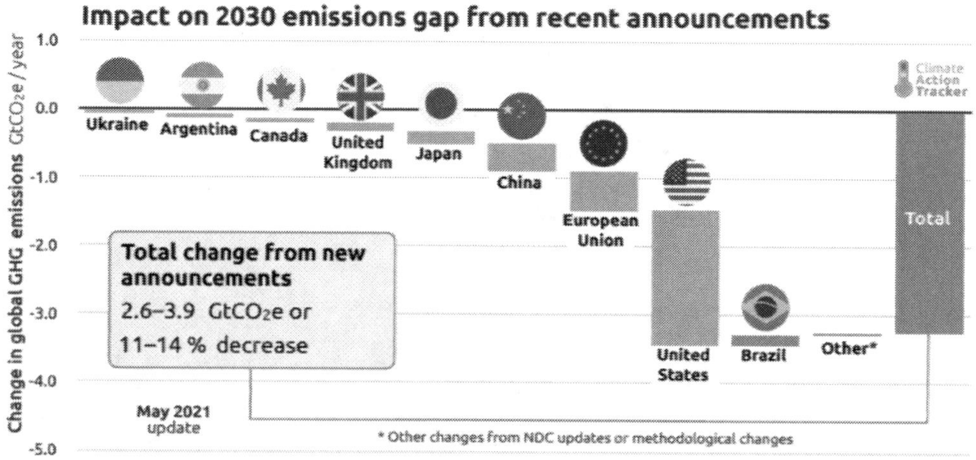

Fig. 2. Impact of recent NDC announcements and submissions since September 2020 on reducing the 2030 emissions gap [3].

We highlighted three key points that are challenging to achieving the updated Nationally Determined Contribution of Ukraine to the Paris Agreement [7].

Political factor. NDC liabilities are determined based on maintaining the transit of natural gas through the territory of Ukraine in the amount of 40-55 billion m^3 per year). However, the possibility of the Nord Stream 2 project completion was not taken into account, as well as the probability of gas supply to European countries bypassing the gas transmission system of Ukraine and gas supply to Hungary and Croatia through the "Turkish Stream." Moreover, it can result firstly - the loss of the possibility of import by reverse; secondly – the risk of termination of gas transit through the GTS of Ukraine upon expiration in 2024 of the agreement on the organization of transit between Naftogaz and Gazprom.

Energy factor. The unpreparedness of Ukraine's economy to abandon fossil fuels in the short term. In addition, the gradual reduction of the share of coal generation in energy requires a proportional transformation of the energy system, finding the optimal ratio between the volume of renewable energy sources and shunting generation, taking into account the threats to national energy security and energy independence.

Financial and economic factors. The lack of domestic investment resources. Characteristic the national circumstances of reducing greenhouse gas emissions in the context of [7] mentioned: considering the significant deterioration of the main assets of large enterprises, modernization of the economy and ensuring sustainable economic development depends on the inflow of investments from both private sources and international financial institutions. According to estimates, the amount of capital investments until 2030, necessary for the implementation of the NDC, is 102 billion euros. Thus, the fulfillment of obligations under NDC occurs against the background of a long-term trend of "narrowing of the primary sources of fixed capital accumulation and shortage of investment resources in the economy [8] and the lack of a sustained long-term upward trend in household incomes. Approaches for redistribution of investments are not yet in place: the internal scheme of emissions trading is in its infancy - at the stage of creating basics of monitoring, reporting, and verification of GHG emissions. As a separate subspecies of securities and a tool for investing in environmental projects, including low-carbon production, Green bonds were introduced only in July 2021. The system of budget support for climate protection projects is also not formed. Therefore, in achieving the goals, Ukraine expects to receive additional climate financing in the form of grants from developed countries and other donors; soft loans; and other financial instruments to create a separate climate fund and increase funding for the decarbonization of the economy through commercial banks and the private sector. The lack of necessary international support puts on the risk of achieving the goal of the NDC.

3 Steps Forward

Failure to adapt to EU initiatives threatens negative consequences shortly. The implementation of the concept of "carbon footprint" and the mechanism of border carbon adjustment may be an obstacle to access Ukrainian goods to European markets, investment, and overall inclusion of the Ukrainian economy into the economic space in the EU.

The conflict of interests of economic growth and transition to the path of low-carbon development requires proper economic and legal tools to balance them and reflect a unified system of measures to decarbonize the economy in the legislation.

The groundwork for effective implementation of obligations under the Renewed National Contribution of Ukraine to the Paris Agreement includes: resolving the issues of determining appropriate economic and legal means of decarbonization of the economy, the formation of practical recommendations aimed at ensuring an integrated approach to the legal regulation of public relations.

The existence of several strategies, plans, and concepts for the development of the energy sector, the current state of legal support for decarbonization of the economy is characterized: by the lack of a legally defined system of strategic planning in the energy sector, inconsistency of energy policy with other public policies (economic, environmental, social), as a consequence, an unsystematic and segmented approach to the legislative regulation of relevant relations.

Legal support for the decarbonization of the economy should be carried out according to the vectors defined by the Paris Agreement: effective and progressive response to the urgent threat of climate change based on the recognition of specific needs and particular circumstances of the participating countries; ensuring the consistency of financial flows with the direction of low-carbon and resistant to climate change development; introduction of a bottom-up approach, which encourages and ensures participation in mitigation of greenhouse gas emissions not only at the national but also at the regional/local level. For defining the main directions of legal support for implementing these tasks at the level of an integrated act of strategic nature, which would simultaneously provide parameters for updating in the context of decarbonization of Economic and Energy Strategies of Ukraine, other policy documents adopted in the development of their provisions. In the acts of strategic planning, it is necessary to establish the sequence of legislative support of their priorities to ensure the connection between the provisions of the updated legislation under the objectives and the consequences of previously adopted acts.

Among the priority areas for the transformation of legal regulation in the context of decarbonization of the economy is the harmonization of the Renewed National Contribution of Ukraine and the Paris Agreement. In particular, the formation of comprehensive economic and legal mechanisms to reduce greenhouse gas emissions, innovation, and investment support for decarbonization of the economy, increasing energy efficiency. Legal support of Ukraine's "green" energy transition soon should be aimed at transforming energy markets in the context of priority development of renewable energy, introducing a system to stimulate demand for the use of "clean energy." In coal generation, special attention should be paid to the modernization of the coal industry in the context of decarbonization and stimulation of alternative options for reducing greenhouse gas emissions, including reducing associated emissions (not from core activities), which account for significant emissions. Given the prospects for bioenergy development, the urgent task is to reorient the legal regulation from stimulating the quantitative growth of bioenergy potential to ensure proper consideration of the relationship with the goals of sustainable development in general and the adopted climate strategy in particular.

The bottom-up, top-down approach introduced by the Paris Agreement to involve local communities in mitigating greenhouse gas emissions is currently being implemented in public initiatives with the support of international financial organizations and is insufficiently represented in current legislation. Also, we should focus on integrating decarbonization objectives into policies at all levels in the horizontal (sectoral programs and strategies) and vertical (local, regional, and macro-regional levels) levels. Moreover, it should form a comprehensive economic and legal mechanism to reduce greenhouse gas emissions, taking into account the peculiarities of territorial development.

4 Conclusions

Decarbonization requires a string of technological revolutions in each of the major emitting sectors. Creating technological revolutions will require different actions in different sectors. In conclusion, we note that further research should focus on developing scientific and methodological approaches to forming adequate legal and economic support for the decarbonization of the economy. The priorities are:

- assess the current state of legal support for the decarbonization of Ukraine's economy, in particular in the context of fulfilling obligations under the Renewed Nationally Defined Contribution of Ukraine to the Paris Agreement;
- to study the world experience of legal support of decarbonization and to determine the ways of its implementation in domestic practice;
- substantiate the priority areas of legal support for the decarbonization of the economy;
- to determine the main directions of economic and legal support of energy sustainability in the conditions of decarbonization;

- to substantiate the legal basis of the "green" energy transition of Ukraine in the context of the requirements of the Paris Agreement and the European Green Agreement;
- identify the features of the bottom-up (top-down) approach to ensure the participation of local communities in mitigating the effects of greenhouse gas emissions at the regional/local level;
- substantiate the economic and legal principles of decarbonization at the regional and local levels.

Based on the presented research results, the following steps will be: designing scientific and methodological recommendations for the adaptation of foreign experience in this area; developing the multidisciplinary mechanisms of legal regulation for decarbonization. It is also planned to identify ways to harmonize the priority areas of legal support for the decarbonization of the economy with the provisions of the Renewed National Defined Contribution of Ukraine to the Paris Agreement. It will also be essential to develop a set of systematically agreed proposals to improve the legislative and regulatory framework governing the decarbonization of the economy.

References

1. Abidoye, B., Felix, J., Kapto, S. & Patterson, L. (2021). *Leaving No One Behind: Impact of COVID-19 on the Sustainable Development Goals (SDGs).* New York, NY and Denver CO: United Nations Development Programme and Frederick S. Pardee Center for International Futures.
2. Paris Agreement FCCC/CP/2015/10/Add.1 (2015) - https://unfccc.int/process/conferences/pastconferences/paris-climate-change-conference-november-2015/paris-agreement
3. Climate Action Tracker Warming Projections Global Update, (May 2021) - https://climateactiontracker.org/documents/853/CAT_2021-05-04_Briefing_Global-Update_Climate-Summit-Momentum.pdf
4. CAT Climate Target Update Tracker. Ukraine (June 31· 2021) https://climateactiontracker.org/climate-target-update-tracker/ukraine/
5. Climate ambition Summit. United Nations, the United Kingdom, and France, in partnership with Chile and Italy (December 2, 2020) - https://www.climateambitionsummit2020.org/ondemand.php
6. *European Green Deal.* COM/2019/640 final (Brussels, 11.12.2019) https://eur-lex.europa.eu/legal-content/EN/TXT/?uri=CELEX%3A52019DC0640
7. Ukraine First NDC (Updated Submission) 31.07.2021 - https://www4.unfccc.int/sites/NDCStaging/pages/Party.aspx?party=UKR
8. Шовкун І.А. Неоіндустріалізація в Україні: чи є макроекономічні передумови та інвестиційний потенціал? *Економіка і прогнозування.* № 4. с.48-69 (2016)

Economic and Environmental Assessment of Environmental Pollution by Construction Companies in Ukraine in the System of Environmental Management in the Post-Pandemic Period

Katerina Zubko[1], *Hanna* Samus[1], Olena Tsyhanenko[1], *Mariia* Troian[2*], and *Marina* Järvis[3]

[1]Konotop Institute Sumy State University, Department of Technology and Management, 41615, Av. Mira, 24, Konotop, Ukraine
[2]Sumy State Makarenko Pedagogical University, Department of Tourism, Hotel and Restaurant Service, 40002, Romenska str., 87, Sumy, Ukraine
[3]Estonian Entrepreneurship University of Applied Sciences, 11415, Suur-Sõjamäe 10a, Tallin, Estonia.

Abstract. The study's primary purpose is to develop an assessment of the implementation of environmental measures in the system of sustainable development. The development of a comprehensive approach to cost evaluation of environmental innovations allowed to estimate the total cost of the construction project together with environmental and economic damage, choose the most optimal environmental measures for housing, agricultural or road construction, more effectively organise the management system of the construction company. The article considers the importance of making environmental managerial decisions in ensuring society's ecological and economic security. The study proved that construction as a branch of material production is one of the main driving forces of the national economy in terms of results and one of the most potent pollutants. The purpose of environmental and economic assessment of losses from the construction industry is to determine the cost-benefit ratio of the construction project in monetary terms with the mandatory comparison of results before and after the implementation of environmental measures. The analysed ecological and economic indicators of the construction industry, the relationship between the volume of construction work by type of construction products and waste generation by type of economic activity allowed to justify the need for approaches to management decisions in environmental management. The study results can help integrate the requirements for managing the impact of production factors on the ecological and economic condition of the country in order to ensure the safe use of natural resources, environmental protection and economic feasibility (profitability) of management decisions in environmental management.
Keywords: environmental measures, Sustainability, evaluation of the results of environmental innovations, environmental and economic losses of the

* Corresponding author: mariia.yu.troian@gmail.com

construction project, sustainable development management, environmental management.

1 Introduction

At the current post-pandemic stage, management and development of the construction industry is a strategic task in connection with its participation in the creation of fixed assets of other sectors of the economy and providing housing for the population. In recent years, Ukraine has changed environmental characteristics, including disturbing the ecological and economic balance, depleting natural resources and increasing the ecological load on the main natural components, in particular, due to the activities of construction companies. Ensuring the profitability of the construction industry, taking into account environmental requirements, requires environmental and economic justification of environmental programs and measures based on indicators of environmental and economic damage, capital and current environmental costs.

Existing methodological developments for assessing environmental and economic losses do not take into account changes in economic relations and property relations in environmental protection, which significantly underestimates the negative impact of anthropogenic activities on the environment. This determines the choice of the topic of the article and its relevance.

2 Literature review

Theoretical issues of economic assessment of the consequences of anthropogenic pollution of the environment were investigated in the works of O. Balatskiy, S. Rogach, L. Melnyk, L. Hens, [1-3], J. Glasson et al. [4]. Problems of ecological and economic assessment of the impact on the environment of specific sectors of the economy are devoted to the work of scientists O. Amosha, E. Khlobystov, O. Prokopenko [5-7] and others [8-12].

A systematic review of the literature in building modelling, lean construction and Sustainability is presented in the study by P. Saieg et al. [13]. As a result, it is shown that the integration of the environmental approach in the design of construction projects provides significant opportunities to reduce the economic and environmental impact of the construction industry.

The authors M. Lieder and A. Rashid, in their study [14], presents the complex structure of the Concept of Circular Economy (CE) and the practical strategy for the implementation of the restorative economy and the environment. The complex structure is a combined view of three main aspects, namely: the environment, resources and economic benefits in the productive sectors of the economy

So it is necessary to investigate the factors of efficiency, the activities of construction companies, the management of which will ensure a high level of competitiveness and economic security.

2 Purpose and research methods

The study's primary purpose is to develop an assessment of the implementation of environmental measures in the system of sustainable development, namely the development of a comprehensive approach to assessing the value of environmental innovations. This assessment will allow us to estimate the total cost of the construction site together with environmental and economic damage, select the most optimal environmental measures for residential, agricultural or road construction, and organise the management system of the construction company more effectively.

Based on the analysis of the latest information characterising the state's economic, social and environmental problems, including analytical data of public authorities, the importance of environmental management decisions in ensuring environmental and economic security of society. The study proved that construction as a branch of material production is one of the main driving forces of the national economy in terms of results and one of the most substantial pollutants. According to the analysis results, the ecological-economic indicators of the construction industry, the ratio between the volume of construction work by type of construction products and waste generation by type of economic activity were evaluated, which justified the need for approaches to decisions in environmental management.

3 Results

The establishment and determination of ecological and economic damage in a market economy solve the problems associated with legal and economic liability for additional costs borne by both individual economic entities and society as a whole in connection with manufactured violations of natural resources. Environment [15].

The priorities of construction companies are compliance with the principles of corporate social responsibility and prevention of environmental and economic losses, as well as making a profit. During construction, the primary sources of pollution were the construction of ditches and trenches, deforestation, soil damage, formation of construction waste dumps, operation of construction machinery and equipment, etc. (Table 1).

Table 1. Waste generation by types of economic activity for 2005-2019 in Ukraine (thousand tons) [16]

Year	Total	Including							Households
		Agriculture, forestry and fisheries	Mining and quarrying	Processing industry	Supply of electricity, gas, steam and air	Water supply; sewerage, waste management	Construction	Other types of economic activity	
2005	2411,8	192,6	32,5	2037,0	10,5	44,7	59,2	35,3	-
2006	2370,9	256,8	26,5	2010,4	22,0	33,5	3,9	17,8	-
2007	2585,2	223,8	19,7	2269,0	9,1	3,2	3,6	56,8	-
2008	2301,2	215,7	16,1	1989,3	8,8	2,7	3,1	65,5	-
2009	1230,3	230,8	16,8	916,6	9,4	4,0	2,5	50,2	-
2010	425914,2	8568,2	321889,8	75950,4	8641,0	1698,7	189,0	2254,7	6722,4
2011	447641,2	12438,2	330301,9	79554,7	9901,2	7253,2	497,9	2517,3	5176,8
2012	450726,8	10238,7	338025,0	78814,8	9812,0	1003,9	519,2	4343,8	7969,4
2013	448117,6	10311,8	341363,2	74279,0	9346,4	1011,2	639,3	2140,5	9026,2
2014[2]	355000,4	8451,4	267506,1	64755,8	5972,7	612,5	131,2	1256,4	6314,3
2015[2]	312267,6	8736,8	232642,4	56506,3	6597,5	594,2	89,9	1047,2	6053,3
2016[2]	295870,1	8715,5	217907,8	53857,9	7511,5	457,4	88,9	984,6	6346,5
2017[2]	366054,0	6188,2	313738,2	32176,7	6191,7	408,7	493,8	998,7	5858,0
2018[2]	352333,9	5968,1	301448,9	31523,2	6322,7	419,6	378,8	1148,7	5543,5
2019[2]	441516,5	6750,5	390563,8	30751,8	5959,2	410,1	188,7	1405,8	5896,7

[1]Until 2010, data on wastes of I-III hazard classes were given.

[2]Excluding the temporarily occupied territory of the Autonomous Republic of Crimea, the city of Sevastopol and part of the anti-terrorist operation zone.

As a rule, the impact of construction sites on ecosystems is complex: construction creates an additional environmental burden and causes deterioration of human health if poor quality building materials and structures are used during the construction of houses.
The world's buildings use about 40% of all primary energy consumption, 67% of electricity, 40% of raw materials and 14% of all drinking water supplies, and produce more than 35% of all carbon dioxide emissions and almost half of all municipal solid waste [17].

The volume of construction work performed by types of construction products1 in Ukraine is presented in table 2.

Table 2. The volume of construction work performed by types of construction products[1] in Ukraine (UAH million) (Statistics, 2010-2019)

Year	Construction, total	Buildings	Including		Engineering structures
			residential	non-residential	
2010	42918,1	19659,1	6876,5	12782,6	23259,0
2011	61671,7	26745,4	8137,1	18608,3	34926,3
2012	62937,2	28104,8	8523,0	19581,8	34832,4
2013	58586,2	28257,3	9953,1	18304,2	30328,9
2014	51108,7	24856,5	11292,4	13564,1	26252,2
2015	57515,0	28907,5	13908,8	14998,7	28607,5
2016	73726,9	38106,4	18012,8	20093,6	35620,5
2017	105682,8	52809,6	23730,0	29079,6	52873,2
2018	141213,1	66791,6	29344,8	37446,8	74421,5
2019	181697,9	83589,3	33208,8	50380,5	98108,6

[1] Excluding the temporarily occupied territory of the Autonomous Republic of Crimea and the city of Sevastopol, in 2014-2019 also without part of the anti-terrorist operation zone.

Thus, quite often, already during the construction of residential and non-residential facilities, construction companies harm the environment: the terrain changes, the vegetation changes, natural plantations are replaced by artificial ones; sometimes newly introduced objects change the conditions of insolation; shading areas, changing the mode of evaporation of moisture.

The concept of rationing the impact of construction on the environment provides:

1) development of a unified system of normative values of maximum permissible levels of construction impact on land, water and atmospheric resources, as well as the generation of construction waste;

2) protection of natural resources from losses caused by the action of this impact on various components of the natural environment;

3) prevent significant functional disturbance of ecosystems due to construction's direct or indirect impact on specific environment components.

In the context of increasing construction and installation work, it is advisable to develop methodological approaches to assessing the damage from the negative impact of construction on the environment and consumers and comparing the benefits of these measures with the costs incurred by the population to restore their health.

The authors believe that it is necessary to provide the following measures to assess the environmental and economic damage due to the negative impact of construction on the population:

- assessment of the number of citizens who fall under the dangerous influence of construction;
- assessment of the probable change in the deterioration of public health;
- quantitative assessment of environmental and economic damage from changes in morbidity among the population;
- assessment of the cost of measures to prevent environmental and economic damage;

- calculation of possible ecological and economic damage;
- analysis of alternative uses of building materials and opportunities to change construction technology.

The analysis of the possible impact of construction on the environment should include the benefits and costs of implementing any construction project. However, there are many difficulties determining the benefits and costs locally, globally, and temporally and spatially. In these cases, the benefit that can be obtained in one region (district of the city) may be accompanied by additional costs incurred in other regions (districts, cities); similarly, it is not beneficial for one region (city district) - it may be vital for another, etc.

The total value of the construction object, together with the ecological and economic damage, should be less than the total benefit from its realisation. This is a necessary condition for evaluating the decisions made in construction. In other cases, the implementation of the housing project should be cancelled.

It is advisable to evaluate the benefits using the following methods: based on rent, opportunity cost, market valuation.

Cost estimation is performed on the following elements:

1. Material costs;
2. Labor costs with accruals;
3. Depreciation deductions;
4. Expenses related to social activities;
5. Costs related to compensation or prevention of environmental and economic damage.

One of the ways to solve this problem is to develop a comprehensive approach to valuing the results of environmental innovations. The task of the environmental component is to reduce the negative impact on the environment, which requires significant capital investment in construction and thus increase the cost of production.

The economic component involves reducing the cost of production by improving the management system of construction companies, saving production resources. At the same time, the impact of the environmental component of the enterprise on its economic activity is ambiguous. Thus, the implementation of capital investments changes not only the cost of construction products but also the overall cost structure of the enterprise. Such structural changes can positively or negatively affect the efficiency of the entire enterprise. The implementation of environmental measures may lead to the receipt (loss) of an additional amount of profit for the enterprise as a whole.

The effect of environmental measures in this approach should be considered as a multiplier effect of accompanying opportunities. After all, the introduction of environmental measures involves changes in production costs, production volumes, net income, fees for using natural resources, taxes on environmental pollution and reducing the cost of paying fines for excessive pollution.

Such changes lead to a change in the cost structure of the enterprise, in which fixed costs may not change, and variables, in turn, may decrease due to cost savings on materials, wages, fuel, electricity and reduce the cost of fines for environmental pollution.

The main result of implementing environmental measures by enterprises of the construction industry is an additional economic result of implementing environmentally-friendly innovations.

An additional accompanying result of the implementation of environmental measures by construction companies is positive or negative for enterprises changes in the cost structure or corresponding changes in the amount of operating leverage as a result of the implementation of environmental measures by construction companies, ie

$$\Delta E_{ol} = \Delta K_{ol} \cdot P_{fp} \qquad (1)$$

where ΔE_{ol} - change in the amount of operating leverage;

ΔK_{ol} - change in the operating leverage ratio as a result of the implementation of environmental measures;

P_{fp} - the amount of profit in the forecast period, UAH

In turn

$$\Delta K_{ol} = K_{ol}^{1} - K_{ol}^{0} \quad (2)$$

Where K^{0}_{ol}, K^{1}_{ol} - respectively, the coefficients of operating leverage in the base and forecast periods.

When calculating the coefficient of operating leverage, it is necessary to determine the ratio of fixed and variable costs, which affects the operating leverage by the formula:

$$K_{ol} = C_f / C_v \quad (3)$$

where K_{ol} is the operating leverage ratio;

C_f - fixed costs for the volume of construction products, UAH;

Cv - variable costs for the volume of construction products, UAH

Therefore, the increase in operating leverage in the forecast period during the implementation of environmental measures by construction companies is determined by the formula:

$$K_{ol}^{1} = K_{cr}^{1} - K_{ol}^{0} / K_{cr}^{0} \quad (4)$$

Thus, the total effect of the implementation of environmental measures E_E can be defined as the sum of direct and accompanying effects, UAH:

$$E_E = P + \Delta E_{ol} \quad (5)$$

where P is the direct effect of the implementation of environmental measures

Thus, taking into account all the potential opportunities for implementing environmental innovations will assess the total economic effect of the construction industry. Only then can we talk about the economic efficiency of the construction industry.

Methodologically, the determination of ecological and economic damage from environmental pollution by a construction company should be carried out in the following sequence:

1. assessment of the level of environmental pollution;
2. identification of the relationship between the environmental pollution level and its qualitative and quantitative impact on the business entity and the territory population;
3. calculation of environmental and economic damage;
4. making changes to the construction project;
5. the formation of strategic behaviour of the construction company concerning its interaction with the environment.

Recent studies show the excess of environmental and economic damage over the environmental tax from enterprises in the construction industry, differentiated by the type of construction (Table 3).

Note that construction production's environmental and economic efficiency involves the supply of competitive goods and services that meet human needs, improve quality of life while reducing the negative impact on the environment and resource consumption throughout the life cycle of these goods and services.

Table 3. Ecological and economic damage from construction and installation work in residential, road and agricultural construction and its relationship with the value of the environmental tax.

Types of construction and installation work	Ecological and economic damage, UAH / year	Ecological tax, UAH / year	+\- %
Housing			
1. Earthworks	7505,68	771,56	
2. Assembly and concrete works	353,16	16,25	
3. Brick works	1444,16	87,69	
4. Interior finishing works	1681,42	77,69	
5. Exterior finishing works	1106,75	47,13	
Total	**12091,20**	**1000,32**	**8,2**
Road construction			
1. Earthworks	4509,72	553,11	
2. Preparation for laying, laying and pumping of asphalt (per 1000 m2)	8828,71	1680,55	
Total	**13338,43**	**2233,66**	**16,7**
Agricultural construction			
1. Earthworks	2952,41	407,73	
2. Assembly and concrete works	686,01	54,73	
3. Brick works	1444,89	37,36	
4. Exterior finishing works	936,26	11,56	
Total	**6019,57**	**511,38**	**8,5**

The urgency of raising the issue of development and implementation of environmental and economic assessments is related to the need to develop the theory of economic efficiency of social production, increase the reliability of economic calculations at the level of management decisions, justification of state program approaches to solving local social and industrial problems.

Ensuring environmental and economic efficiency in the construction industry requires a systematic approach to management decisions. The optimisation of management decisions in ecology should be considered in close cooperation with nature management, ecology and economy. The subsequent management decisions are possible due to the simultaneous consideration:

a) the results of basic and applied research on the properties of the building materials themselves;

b) new technologies of secondary use of waste, aimed at increasing productivity and reducing the negative environmental impact of construction companies;

c) new organisational decisions on the certification procedure and appropriate decision-making criteria when planning the activities of enterprises in the construction industry.

Competitiveness of the construction enterprise is provided, including its resource component and the realisation price of habitation. Therefore, in our opinion, the most crucial thing will be to focus on maintaining a balance between the economic activity of economic entities in the construction industry and the ecological state of the environment.

An environmentally friendly approach, based on the optimisation of measures to increase the level of environmental friendliness through innovative technologies in construction to preserve the environment and natural resources, should be used to enhance the development of the construction industry.

The results of these actions are to obtain economical, social and environmental effects. The following measures can be proposed to stimulate the introduction of ecological construction in Ukraine:

1. economic incentives from the state;
2. conducting the state ecological examination of new construction projects as, in fact, the only legitimate tool for recognising the ecological capacity of future facilities;
3. intensification of environmentally oriented investment activities;
4. development of the system of ecological education;
5. orientation of the state on resource-saving programs;
6. taking into account innovative aspects in ensuring the environmental safety of economic development;
7. formation of the market demand for environmentally friendly products.

Implementing environmental measures on a systemic and integrated basis is associated with the need to maintain a balance between production and the state of the environment.

4 Conclusions and prospects for further research

Sustainable development of the construction industry depends on the possibility of implementing measures to protect the environment. The economic efficiency of using natural resources is formed under the influence of minimising environmental risks that directly impact ecosystems.

Considering all the potential opportunities for implementing environmental innovations will assess the total economic effect and determine the cost structure of enterprises in the construction industry. Furthermore, one of the tools for managing construction companies should be to manage their greening.

References

1. L. Hens, L. Melnik, E. Boon, *Environmental Economics* (Kyiv, Naukova Dumka, 1998)
2. S. Rogach, Prace z zakresu nauk spolecznych, 14 (2013)
3. L. G. Melnyk, I. Degtyarova, J. Economics of Ukraine, **5**, 75-83 (2011) (in Ukrainian)
4. J. Glasson, R. Therivel, Introduction to Environmental Impact Assessment (Routledge, 2019), https://doi.org/10.4324/9780429470738
5. O. I. Amosha, Man and the environment: economic problems of ecological safety of production (Institute of Industrial Economics, Kyiv, Scientific Opinion, 2002) (in Ukrainian)
6. L.O. Dobryanska, L.V. Zharova, E.V. Khlobistov, Strategic potential of ecological safety: technology of economic growth (Lviv, Ukrainian bestseller, 2012) (in Ukrainian)
7. O. Prokopenko, I. Marekha, MIND, **1**, (2016) https://mindjournal.wseh.pl/sites/default/files/article/09-16/framing_ecological_values_in_the_global_natural_resource_economics.pdf
8. K. Zubko, Economy and society, **9**, 6826–832 (2017), URL: http://economyandsociety.in.ua/journal/9_ukr/9_2017.pdf (in Ukrainian)
9. S.Z. Vesperis, K.Yu. Zubko, G.I. Samus, Priazovsky Economic Bulletin, **1 (18)** (2020) URL: http://pev.kpu.zp.ua/vypusk-18 (in Ukrainian)
10. M. D. Domashenko, O. V. Panchenko, M. Yu. Troyan, T. Yu Makarenko. The International Journal of Ecology and development, **35(2)**, 31-37 (2020). http://www.ceser.in/ceserp/index.php/ijed/article/view/6450
11. M. Yu. Troian, S.M. Frolov, Marketing and Management of Innovation, **3 (2)**, 125-134 (2011), URL: https://mmi.fem.sumdu.edu.ua/journals/2011/3_2/125-134 (in Ukrainian)

12. L. Hens, L. Melnyk, O. Matsenko, O. Chygryn, and C.C. Gonzales, Marketing and Management of Innovations, **3**, 272-284. http://doi.org/10.21272/mmi.2019.3-21, (2019)
13. P. Saieg, E. D. Sotelino, D. Nascimento, R. G. Gusmão Caiado, Journal of Cleaner Production, **174**, 788-806 (2018), https://doi.org/10.1016/j.jclepro.2017.11.030
14. Michael Lieder, Amir Rashid, Journal of Cleaner Production, **115**, 36-51 (2016), https://doi.org/10.1016/j.jclepro.2015.12.042
15. M.Kyslyy, Bulletin of SSU, Series "Economics", **1(2)**, 43-50 (2007) (in Russian)
16. State Statistic Service of Ukraine, http://www.ukrstat.gov.ua/
17. Z. A. Gaevskaya, Yu. S. Lazareva, A. N. Lazarev, Young scientist, 16(96) (2015) https://moluch.ru/archive/96/21620/ (in Ukrainian)

Ensuring the Competitiveness of the Ukrainian Woodworking Industry in the Post-Pandemic Period

Irina Hubarieva[1*], *Oleksandr* Hubariev[2], *Volodymyr* Zinchenko[3], *and Pavlo* Pronoza[4]

[1] National Academy of Sciences of Ukraine, Research Centre for Industrial Problems of Development, 1a Inzhenernyi Ln., Kharkiv, 61166, Ukraine
[2] Kharkiv State Academy of Culture, 4 Bursatskyi Des., Kharkiv, 61057, Ukraine
[3]National Academy of Sciences of Ukraine, Research Centre for Industrial Problems of Development, 1a Inzhenernyi Ln., Kharkiv, 61166, Ukraine
[4]Simon Kuznets Kharkiv National University of Economics, Nauky avenue, 9-A, Kharkiv, 61166, Ukraine

Abstract. The woodworking industry is one of the most critical components of the forest sector in Ukraine and world countries. This paper aims to develop directions for ensuring the competitiveness of the woodworking industry in the post-pandemic period. The study resulted in the elaboration of the theoretical framework for identifying the main substantive determinants of industrial competitiveness in world countries through bibliometric analysis using VOSviewer v.1.6.10. The conducted cluster analysis provided for the identification of 4 clusters of scientific research dealing with issues of industrial competitiveness (Cluster 1 is focused on the formation of a competitive business strategy for the development of industrial enterprises; Cluster 2 – on the development of the export-import potential of industrial enterprises; Cluster 3 – on the identification of relationships between ensuring industrial competitiveness and globalisation processes; Cluster 4 – on the implementation of Industry 4.0 and innovative technologies). The obtained theoretical conclusions and generalisations became the basis for forming a system of indicators for assessing the competitiveness of economic activities in the country's woodworking industry. An integrated assessment of the competitiveness of the woodworking industry in Ukraine and world countries was carried out. Using cluster analysis (k-means), world countries, including Ukraine, were clustered according to the value of composite indicators for assessing the competitiveness of individual economic activities in the woodworking industry. Ukraine fell into the cluster characterised by the high competitiveness of individual economic activities in the woodworking industry. Measures were formulated to ensure the competitiveness of the Ukrainian woodworking industry in the post-pandemic period. Keywords: competitiveness, woodworking industry, a composite indicator, cluster analysis, bibliometric analysis, post-pandemic period

* Corresponding author: gubarievairyna@gmail.com

1 Introduction

The COVID-19 pandemic has affected the lives and health of people and the movement of goods, labour, consumption, transportation, and distribution of resources. The problem of ensuring the competitiveness of the industrial sector is becoming increasingly pressing in the post-pandemic period. The woodworking industry is one of the essential components of the forest sector in Ukraine and world countries: it includes enterprises engaged in mechanical and chemomechanical pulping, wood processing and is a critical element of the technological chain: forest cultivation–logging–processing–manufacturing of final products.

The competitiveness of the industrial sector forms the competitiveness of the national economy. The woodworking industry of any country across the world includes economic activities that differ in terms of complexity of production technologies, gross value added, and efficiency both for the country's economy as a whole and individual business entities.

2 Analysis of current research of the problem

The urgency of ensuring the competitiveness of objects at different levels of the hierarchy is evidenced by the frequency of citation of publications in this field in the Scopus scientometric database. The top five most cited publications on the problem of competitiveness, indexed in the Scopus database during 1990–2021, includes the following works: D. Teece (2007) [1] – 5050 citations, M. Porter (2000) [2] – 2236 citations, P. Bansal, K. Roth (2000) [3] – 1931 citations, M. Porter (1991) [4] – 1786 citations, J. Jansen et al. (2006) [5] – 1656 citations.

"Competitiveness is one of the main categories widely used in the theory and practice of economic sciences, a multi-faceted concept translated from Latin as rivalry, struggle to achieve the best results" [6].

In a global economy, competitiveness can be defined as the ability of the country's industrial sector to meet international competition in its market and markets of other countries. Therefore, it is relevant to develop analytical tools for assessing the level of competitiveness of industries in Ukraine and world countries.

The problem of ensuring industrial competitiveness has attracted the attention of theorists and practitioners for a long time. This study aims to systematise and generalise the existing scientific publications, indexed in the Scopus and Web of Science scientometric databases, on the competitiveness of the industry as a whole and the competitiveness of the woodworking industry in particular.

The preliminary analysis of scientific publications on industrial competitiveness, indexed in Scopus and Web of Science, showed that the oldest Scopus indexed article on this issue is the work by R. Stuchtler (1967) [7], which is focused on the investigation of the competitiveness of the German metallurgical industry. The oldest article on the competitiveness of the woodworking industry is the work by V. Carlsson, L. Ohlsson (1976) [8], which identifies the structural determinants of Sweden's export competitiveness: the availability of domestic timber in combination with high capital intensity and investment in the woodworking industry.

However, one of the latest publications on industrial competitiveness is F. Suarez et al. (2022) [9], where scientists note the need to use clustering to ensure the textile industry's competitiveness in the post-pandemic era. One of the most recent publications on the competitiveness of the woodworking industry is the article by S. Krišt'Aková et al. (2018) [10], which deals with assessing the efficiency of woodworking enterprises in Slovakia and Bulgaria.

Among the publications indexed in Web of Science, the oldest document on the problem of industrial competitiveness is the article by CS Giesecke [11] (1972), where the author investigates costs associated with competitiveness in the Swedish industry; and the oldest document on the problem of competitiveness of the woodworking industry is the article by H. Kato (1974) [12], which deals with the assessment of the competitiveness of woodworking machines.

One of the recent publications on industrial competitiveness, indexed in Web of Science, is the article by L. Gastaldi et al. (2022) [13], which deals with studying the consequences of the industry's transition to Industry 4.0. One of the recent publications on the problem of competitiveness of the woodworking industry is that by J. Michal et al. (2021) [14], which proposes a model for sustainable development of woodworking enterprises in the Czech Republic.

To identify the contextual and temporal patterns of the representation of research on industrial competitiveness in the scientific literature, a bibliometric analysis of scientific publications indexed in the Scopus and Web of Sciences databases was carried out using VOSviewer v. 1.6.10 [15].

In the study, 24183 scientific publications indexed in Scopus and 29903 indexed in Web of Science for the period 1971-2021 were analysed. The search was performed using the keywords "competitiveness, industry".

Considering the results of the bibliometric analysis, we can single out 4 contextual clusters of research dealing with industrial competitiveness, which include the most significant number of publications: Cluster 1, which contains publications that consider enterprise competitiveness (the main keywords: business, competitive advantages, product, resources, strategy, management); Cluster 2, which is formed by works focused on identifying bilateral relationships between industrial competitiveness and market determinants (main keywords: export, market size, monopoly, protectionism, taxes); Cluster 3, includes publications that deal with revealing the impact of globalisation processes, introduction of innovations, increase in labor productivity and level of qualification of workers, infrastructure development on industrial competitiveness; Cluster 4, covers articles on industrial competitiveness in the context of government regulation, industrial policy, clusters, technological development, Industry 4.0.

3 Key issues and hypothesis of the research

Given the conducted analysis of publications considering problems of ensuring industrial competitiveness, in our opinion, further research is required on several issues, in particular:

- how the level of competitiveness of individual economic activities affects the level of competitiveness of the country's woodworking industry;
- what patterns of development of the woodworking industry can be distinguished in the clusters formed by world countries;
- what the directions for ensuring the competitiveness of the woodworking industry in the post-pandemic period are.

This study presents an attempt to solve these problems; its central hypothesis is formulated as follows: ensuring the competitiveness of the Ukrainian woodworking industry in the post-pandemic period depends on the development of the economic activities in the woodworking industry that are export-oriented and have the most negligible dependence of intermediate production on imports.

4 Materials and Methods

The research aims to develop directions for ensuring the competitiveness of the woodworking industry in the post-pandemic period.

The object of the research is the competitiveness of the woodworking industry in world countries and Ukraine.

To achieve the set goals, the following research objectives are set:

1. to elaborate a theoretical framework for assessing and analysing the level of industrial competitiveness in world countries;
2. to assess the level of competitiveness of individual economic activities in the woodworking industry in world countries and Ukraine (using the composite indicator);
3. to carry out the clustering of world countries and Ukraine according to the level of competitiveness of the woodworking industry and identify groups of countries that are homogeneous in the level and are characterised by similar patterns of development;
4. to identify ways for ensuring the competitiveness of the woodworking industry in the post-pandemic period.

The primary sample of research on competitiveness of economic activities in the woodworking industry in world countries comprises data of analytical reports on the world's forest resources of the Food and Agriculture Organization of the United Nations (FAO) and the United Nations Economic Commission for Europe (ECE) [16-24].

Given the tasks set, the study employs the following methods:

1. general scientific methods: analysis and synthesis, generalisation, and bibliometric analysis, to elaborate the theoretical framework for ensuring and assessing the competitiveness of the woodworking industry;
2. the method for calculating the composite indicator, to assess the competitiveness of economic activities in the woodworking industry in world countries and Ukraine;
3. cluster analysis, to cluster world countries and Ukraine by the level of competitiveness of the woodworking industry.

Thus, the study consists of three main parts.

The first one deals with elaborating the theoretical framework of the study. It involves the analysis of research works on the issues under study and is aimed at the substantiation, on this basis, of approaches to ensuring and assessing the competitiveness of the country's woodworking industry.

The second part focuses on calculating the composite indicator of the competitiveness of economic activities in the country's woodworking industry. As a basis for assessing the competitiveness of economic activities in the country's woodworking industry, it is proposed to put a hierarchically built system of indicators, including the composite indicator and sub-indicators (grouped by corresponding components). The composite indicator is compiled using the taxonomy method, which makes it possible to determine the level of and disproportions in the development of the country's woodworking industry and can serve as a basis for the formation of directions for increasing its competitiveness.

The third part of the study deals with clustering world countries according to the level of competitiveness of economic activities in the woodworking industry and determining ways to ensure the competitiveness of the woodworking industry in Ukraine in the post-pandemic period. The primary research method used for this purpose is cluster analysis.

Clustering is the most common tool for object typology. It implies classifying objects (in our study, countries) into relatively homogeneous groups, considering the considered number of features (indicators, variables). Cluster analysis methods are covered in sufficient detail in many scientific works and are widely used in economic research. When clustering countries by level of competitiveness of the woodworking industry, the Euclidean distance was chosen to measure the distance between objects, and the Ward principle – to measure the distance between clusters. The result of applying cluster analysis in the framework of this study should be the identification of groups of countries that are homogeneous in terms of competitiveness of the woodworking industry and have similar patterns of development.

5 Results

Summarising the results of the study of the modern scientific literature, the main substantive determinants of competitiveness of the country's industry were identified through bibliometric analysis using VOSviewer v.1.6.10. The formalisation of contextual features of the studied concept is based on the bibliometric analysis results of 24183 scientific publications indexed in the Scopus scientometric database and 29903 ones indexed in Web of Science for the period 1971-2021, was carried out. This made it possible to find out that analysis of relationships between the competitiveness of the industry and an enhancement in the efficiency and sustainable development of industrial enterprises, introduction of innovative technologies, greening of production is becoming increasingly popular.

Thus, we identified four clusters of scientific research dealing with issues of industrial competitiveness (Cluster 1 is focused on the formation of a competitive business strategy for the development of industrial enterprises; Cluster 2 – on the development of the export-import potential of industrial enterprises; Cluster 3 – on the identification of relationships between ensuring industrial competitiveness and globalisation processes; Cluster 4 – on the implementation of Industry 4.0 and innovative technologies). The obtained theoretical conclusions and generalisations provided the basis for forming a system of indicators for assessing the competitiveness of economic activities in the country's woodworking industry.

Industry of any country across the world includes economic activities that differ in terms of complexity of production technologies, gross value-added, and efficiency both for the country's economy as a whole and individual business entities.

The manufacturing of wood and wood products is divided into the preparation and primary processing of wood and secondary wood. Preparation and primary processing of wood include the following: production of veneer, pulp chips and timber (planks, beams, rough-sawn timber). Secondary processing involves the following: production of plywood, boards (chipboard, fiberboard, MDF, OSB, etc.), trim mouldings, laminated wood beams and bars, laminated wood constructions [25; 26].

The study shows that the level of development and efficiency of the country's woodworking industry is determined by three economic activities: timber production, veneer production, and board materials.

Let us assess the competitiveness of individual economic activities in the woodworking industry in Ukraine and world countries.

Based on the results of the bibliometric analysis and data of the annual analytical reports on the world's forest resources of the Food and Agriculture Organization of the United Nations (FAO) and the United Nations Economic Commission for Europe (ECE) [7-16], the following system of sub-indicators is proposed to assess the competitiveness of economic activities in the woodworking industry (Tbl. 1).

It is proposed to carry out an integrated assessment of the competitiveness of economic activities in the country's woodworking industry (I_{WI}) using the formula:

$$I_{WI} = \sum_{z=1}^{3} a_z K_{zj} \qquad (1)$$

where a_z is the coefficient of variation of the composite indicator for assessing the competitiveness of the z-th economic activity in the woodworking industry;

3 is the number of economic activities in the country's woodworking industry;

C_{zj} is the composite indicator for assessing the competitiveness of the z-th type of economic activity in the wood-processing industry in the j-th country, calculated by the formula:

$$C_{zj} = \sum_{l=1}^{3} ß_{lz} X_{lz_j} \qquad (2)$$

where $ß_{lz}$ is the coefficient of variation of the *l*-th sub-indicator for assessing the competitiveness of the z-th economic activity in the country's woodworking industry;

X_{lz_j} is the *l*-th sub-indicator for assessing the competitiveness of the z-th economic activity in the woodworking industry;

3 is the number of sub-indicators for assessing the competitiveness of individual economic activity in the country's woodworking industry.

Table 1. The system of sub-indicators for assessing the competitiveness of economic activities in the country's woodworking industry

Sub-indicator	Calculation formula
The share of production of economic activities in the overall production of the woodworking industry	$X_1 = \frac{V_i}{V_{WI}}$ where V_i is the volume of production of economic activities in the woodworking industry; V_{WI} is the overall production of the woodworking industry
The share of exports in the volume of production of economic activities in the woodworking industry	$X_2 = \frac{E_i}{V_i}$ where E_i is the volume of exports of economic activities in the woodworking industry
Import dependence of the economic activities in the woodworking industry	$X_3 = \frac{I_i}{D_i}$ where I_i is the volume of imports of economic activities in the woodworking industry; D_i is the volume of domestic consumption of economic activities in the woodworking industry

Figure 2 presents a diagram of the distribution of Ukraine and world countries according to the composite indicator for assessing the competitiveness of economic activities in the woodworking industry in 2019.

As can be seen from Figure 1, the leading countries in terms of competitiveness of economic activities in the woodworking industry in 2019 are Latvia (0.4906), Austria (0.4255), Ukraine (0.4144), Russia (0.4096), and Estonia (0.3969).

The group of countries with a low level of the composite indicator for assessing the competitiveness of economic activities in the woodworking industry includes Belgium (0.2724), Sweden (0.2527), Japan (0.2508), Greece (0, 2237), and Great Britain (0.2165).

Using cluster analysis (k-means), countries of the world, including Ukraine, are classified according to the value of the combined indicators for assessing the competitiveness of economic activities in the woodworking industry.

Cluster 1, which is characterised by a high level of competitiveness of individual economic activities in the woodworking industry, comprises fifteen countries, namely: Austria, Brazil, Denmark, Estonia, Canada, Latvia, Lithuania, Mexico, Portugal, Russia, Romania, Hungary, Finland, the Czech Republic, and Ukraine.

Cluster 2, characterised by a low level of competitiveness of individual economic activities in the woodworking industry, is formed by 21 countries: Australia, Belgium, Bulgaria, Great Britain, Greece, India, Ireland, Spain, Italy, China, Korea, Netherlands, Germany, Poland, Slovakia, Slovenia, USA, Turkey, France, Sweden, and Japan.

The conducted analysis has made it possible to recommend the countries that belong to the 1st cluster (including Ukraine), characterised by a high level of competitiveness of economic activities in the woodworking industry, to focus on increasing the export of high value-added wood products, such as board materials.

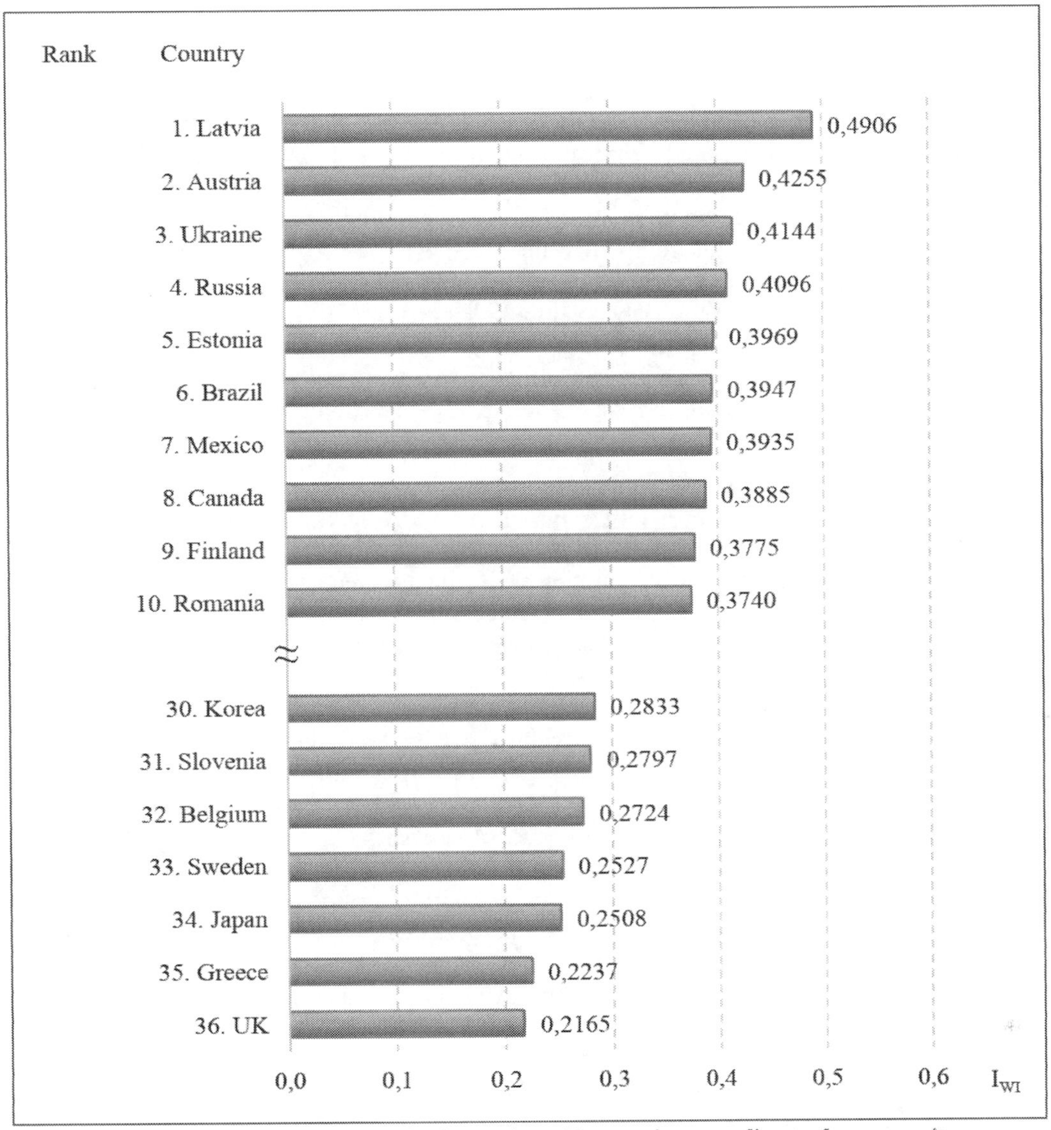

Fig 1. Diagram of the distribution of Ukraine and world countries according to the composite indicator for assessing the competitiveness of economic activities in the woodworking industry in 2019 (developed by the authors)

The clustering of the studied countries by individual economic activities in the woodworking industry demonstrates that Ukraine predominantly falls into the cluster characterised by a high share of production of the economic activities and low values of the share of exports and import dependence of the sub-indicators for assessing the competitiveness of economic activities in the woodworking industry.

In the post-pandemic period, the issues of creating conditions for preventing and minimising economic losses and recession in the industrial sector of Ukraine are becoming aggravated.

During the COVID-19 pandemic, in Ukraine, enterprises in various industries temporarily, fully or partially ceased their operation, which was due to imposing strict quarantine rules, a decrease in the demand for non-essential goods, the introduction of restrictions on Ukrainian products in the markets of world countries, etc. This resulted in the loss of income, potential loss of specialists by these entities and even their bankruptcy.

State support for woodworking enterprises in the post-pandemic period should include the following measures: developing strategic directions for the growth of the woodworking industry; relieving the tax and administrative pressure; decreasing the share of the shadow economy in the forest sector by providing incentives for the holding of timber tenders via the Prozoro electronic procurement system; under the deteriorating conditions in foreign markets, it is crucial to stimulate industrial production by increasing domestic demand through government procurement and protection of local producers.

The measures to ensure the competitiveness of the woodworking industry in Ukraine in the post-pandemic period also include: enhancing secondary processing by reorienting the industry towards manufacturing higher value-added products (e.g., boards), with the focus on increasing sales in the domestic and foreign markets; stimulating the design and construction of infrastructure and industrial parks to attract investment; implementing the ecological modernisation of enterprises; encouraging the use of digital innovation; expanding e-commerce.

4 Conclusions and Discussion

Thus, the study has provided for formulating the following main conclusions and ideas:

1. As a result of the study, the theoretical framework for identifying the main substantive determinants of industrial competitiveness in world countries has been elaborated (through bibliometric analysis with the use of VOSviewer v.1.6.10). The formalisation of contextual features of the studied concept is based on the bibliometric analysis results of 24183 scientific publications indexed in the Scopus scientometric database and 29903 ones indexed in Web of Science for the period 1971-2021, was carried out. This made it possible to find out that analysis of relationships between the competitiveness of the industry and an enhancement in the efficiency and sustainable development of industrial enterprises, introduction of innovative technologies, greening of production is becoming increasingly popular. Thus, we identified 4 clusters of scientific research dealing with issues of industrial competitiveness (Cluster 1 is focused on the formation of a competitive business strategy for the development of industrial enterprises; Cluster 2 – on the development of the export-import potential of industrial enterprises; Cluster 3 – on the identification of relationships between ensuring industrial competitiveness and globalisation processes; Cluster 4 – on the implementation of Industry 4.0 and innovative technologies). The obtained theoretical conclusions and generalisations provided the basis for forming a system of indicators for assessing the competitiveness of economic activities in the country's woodworking industry.
2. An integrated assessment of the competitiveness of individual economic activities in the woodworking industry in Ukraine and world countries has been carried out. It is determined that the leading countries in terms of the competitiveness of the woodworking industry in the considered period are: Latvia (0.4906), Austria (0.4255), Ukraine (0.4144), Russia (0.4096) and Estonia (0.3969). The group of world countries with a low level of the composite indicator for assessing the competitiveness of the woodworking industry includes Belgium (0.2724), Sweden (0.2527), Japan (0.2508), Greece (0.2237), and the UK (0.2165).
3. Using cluster analysis (k-means), world countries, including Ukraine, have been classified according to the value of the composite indicators for assessing the competitiveness of individual economic activities in the woodworking industry.

Cluster 1, characterised by a high level of competitiveness of individual economic activities in the woodworking industry, includes fifteen countries of the world, namely: Austria, Brazil, Denmark, Estonia, Canada, Latvia, Lithuania, Mexico, Portugal, Russia, Romania, Hungary, Finland, the Czech Republic, and Ukraine. Cluster 2, characterised by a low level of competitiveness of individual economic activities in the woodworking industry, comprises 21 countries: Australia, Belgium, Bulgaria, Great Britain, Greece, India, Ireland,

Spain, Italy, China, Korea, Netherlands, Germany, Poland, Slovakia, Slovenia, the USA, Turkey, France, Sweden, and Japan.

4. It has been established that a similar situation is observed with the competitiveness of individual economic activities in the Ukrainian woodworking industry. The clustering of the studied countries demonstrates that Ukraine predominantly falls into the cluster characterised by a high share of production of the economic activities and low values of the share of exports and import dependence of the sub-indicators for assessing the competitiveness of economic activities in the woodworking industry.

5. The methodological approach to assessing the competitiveness of economic activities in the country's woodworking industry has been proposed. The approach considers the scale of production, export orientation, and dependence of intermediate production on imports, making it possible to determine the priority directions for the development of the Ukrainian woodworking industry in the post-pandemic period.

6. Measures to ensure the competitiveness of the woodworking industry in the post-pandemic period have been formulated.

References

1. D.J. Teece Strategic Manage J. **28** (2007)
2. M.E. Porter Econ. Dev. Q. **14** (2000)
3. P. Bansal, K. Roth, Acad. Manage J. **43** (2000)
4. M.E. Porter Strategic Manage J. **12** (1991)
5. J.J.P. Jansen, F.A.J Van Den Bosch, H.W. Volberda, Manage Sci. **52** (2006)
6. S.M. Klimenko, T.V. Omelyanenko, D.O. Barabas and others. Enterprise competitiveness management (2008)
7. R. Stuchtey R. Inter. Econ. **2** (1967)
8. B. Carlsson, L. Ohlsson. Eur. Econom. Rev. **7** (1976)
9. F.R.L. Suárez, O. Flor, C. Chimbo C. Smart Innov. Syst. Technol. **252** (2021)
10. S. Krišt'Aková, N. Neykov, P. Antov, M. Sedliačiková, R. Reh, A.-F. Halalisan, I. Hajdúchová. *Forests*. 12 (2021)
11. CS. Giesecke. Jernkontoret. ANN. 156 (1972)
12. H. Kato. Bus. Japan. 19 (1974)
13. L. Gastaldi, S. Lessanibahri, G. Tedaldi, G. Miragliotta Technol. Forecast. Soc. 174 (2021)
14. J. Michal, D. Brezina, D. Safarik, R. Baduka. Forests. 12 (2021)
15. VOSviewer
16. Forest Products Annual Market Review 2019-2020 (2021)
17. Yearbook of forest products 2019 / FAO (2021).
18. Global Forest Resources Assessments. Country Reports 2020 (2021)
19. Forest Products Annual Market Review 2018-2019 (2020)
20. Forestry Production and Trade (2021)
21. The Global Forest Resources Assessments (2020)
22. The State of the World's Forests (2020)
23. Forest Products Annual Market Review 2018-2019 / Food and Agriculture Organization of the United Nations (2020)
24. Forest Products Annual Market Review 2017-2018 / Food and Agriculture Organization of the United Nations (2019)
25. M. Kyzym, I. Yaroshenko, V. Khaustova, I. Hubarieva I. Formation of strategic priorities for the development of the timber industry complex of Ukraine (2019)

26. I. Gryshova, M. Kyzym, V. Khaustova, V. Korneev, H. Kramarev. Sustainability **12** (2020)

Industries' Bioeconomic Transformation as the Mechanism of Sustainable Development

Victoria Vostriakova[1], *Olena* Rubanenko[2*], *Iryna* Hunko[3], and *Sree Lakshmi* Gundebommu[4]

[1]Department of admimistrative management, Vinnytsia National Agrarian University, Vinnitsya, Sonyachna 3, 21000, Ukraine
[2]Department of Electric Stations and System, Vinnytsia National Technical University, Khmelnytsky highway 95, 21021, Vinnytsya, Ukraine
[3]Department of Electric Stations and System, Vinnytsia National Technical University, Khmelnytsky highway 95, 21021, Vinnytsya, Ukraine
[4]EEE Department, CVR College of Engineering, Mangalpalli, Hyderabad, Ibrahimpatnam, Rangareddy, Telangana 501 510

Abstract. The bioeconomic strategy of Ukraine has not been yet formed, the development of biotechnology is slow, the use of biotechnology is still very fragmentary. The main economic biotechnology sectors of Ukraine are agriculture, pharmaceuticals, food industry and bioenergy. A circular, sustainable bio-economy can become the primary tool for Ukraine's integration into the EU Green Economy, making its economy more sustainable and competitive. The article aims to develop scientific and methodical aspects of the bioeconomic transformation of the national economy of Ukraine. The paper conducts a comprehensive assessment of the bioeconomic potential of the agri-food sector of the Ukrainian economy and identifies the main development priorities. The dynamics of alternative energy consumption and production from alternative energy sources have been provided. Keywords: bioeconomic strategy, the agri-food sector, sustainable bio-economy, biotechnology.

1 Introduction

The bioeconomic transformation involves the main sectors of the national economy of Ukraine to more innovative, resource-efficient, and competitive technology, primarily due to a number of global problems inherent in the current state of Ukraine's economy: environmental degradation of natural resources, global food crisis, inefficient use of resources and energy, waste generation, social inequality, and economic backwardness. All these factors create uncertainty about the future sustainable development of the country.

* Corresponding author: olenarubanenko@ukr.net

World scientific and technological progress and the inconsistency of the economic model of Ukraine's management with the main development trends of the European space form a number of unresolved issues related to ensuring the sustainable development of the country. Transformation of the national economy in accordance with the principles and goals of sustainable development can be ensured by intensifying the objective process of bioeconomy development. The need and objectivity of the process of bioeconomic transformation are presented in this paper.

2 Bioeconomy potential of Ukraine

Agri-food sector development around the world is accompanied by a large number of by-products and waste. Such waste streams are only partially involved in value-adding at various levels (biofuel, fertilizers and animal feed production, composting), while the main volumes generate waste and become a source of environmental danger, which affects sustainable development negatively [1]. This problem is especially actual for Ukraine as a granary of Europe. Ukraine's agriculture plays a central role in the national economy, as it generates more than 20% of GDP [2] and provides more than 40% of export revenue [3]. Ukraine has significant natural and economic potential for further growth and may become a key player in global food security in the coming decades. So, it is necessary to protect the country from environmental challenges arising from climate change. Climate change is a global problem that all countries must address, but it is also a national and economic priority in the case of Ukraine. According to Central Observatory of Ukraine data, 2020 was the hottest for all history [4]. According to the Ministry of Environmental Protection and Natural Resources of Ukraine [5], the country lost 570 thousand hectares of winter crops due to prolonged droughts and extremely severe spring frosts. Subsequent droughts in the spring and summer seasons have led to additional losses, including 200,000 hectares of corn and declining yields across the country.

This situation poses a threat to Ukraine as one of the world's leading exporters of agricultural products. In the 2019-2020 marketing year, Ukraine shipped 57 million to international markets tons of grain, which is about 16% of world exports. However, the previous period's 2020-2021 marketing year exports decreased sharply (by 16%) due to reduced yields caused by abnormal drought [6].

Over the past few years, Ukraine has been ranked among the top five agricultural exporters to the EU. In 2019, the country ranked second among suppliers of organic products to EU consumers. Experts estimate that Ukraine has the potential to achieve a tenfold increase in organic food production in the coming years, with lower yields offset by higher market prices. The fast-growing segment of organic products is just one of many areas where Ukraine can expand its presence in EU agricultural markets. This makes compliance with the European Union's environmental and food standards a long-term need for the country's entire agricultural sector. The European Green Agreement and the EU's "From Farm to Fork" strategy provides Ukraine with additional motivation for a gradual transition to the principles of sustainable development in the agricultural sector.

In 2012, the European Union adopted the strategy of "Innovation for Sustainable Development is Bioeconomy for Europe". Thus, the European Union tried to protect itself from a number of threats faced by Europe and the world at that time [7]. In 2017, this strategy was updated, and a new document called "Review of the European Bioeconomy Strategy 2012" was developed based on the current state of the economy. The primary purpose of this document is to transform countries' economies into more innovative, resource-efficient and competitive, which should ensure the country's food security with the conservation and efficient use of renewable resources, including the preservation of the natural environment [8].

The EU's bioeconomy strategy for 2018 aims to develop a circular, sustainable bioeconomy for Europe, strengthening the link between the economy, society and the environment. It addresses global challenges such as the achievement of the United Nations Sustainable Development Goals (SDGs) and the climate goals of the Paris Agreement. Today, the bioeconomy is the only system that provides the population with food, animals which food base and provides many ecosystem processes, so it is indispensable. As an agrarian country, Ukraine is sufficiently provided with safe and affordable biomass, which, both in the medium and long term, provides an opportunity to meet the needs of the bioeconomic transformation of the national economy. According to some studies, the potential of biomass in Ukraine is so significant that it can provide all sectors of the national economy, and probably outside it, as well as return organic carbon back to the soil.

Ukraine has signed the Paris Agreement, so increasing the share of renewable energy sources is crucial for Ukraine, and the potential and economic benefits of alternative energy development are significant. Bioenergy in Ukraine accounts for the largest share of the total primary energy supply(PES) from renewable energy sources (RES), with approximately 1.9 million tons of energy (59.3%). However, according to official statistics in Ukraine in the field of bioenergy is accounted for only 1.6% of General Electricity Supply (GES); in practice, the renewable energy sector accounted for only 2.8% of Ukrainian GES in 2013 [9]. At the same time, the sale of 20-30% of the 120 million tons of biomass raw materials (including crop, waste, livestock, wood and food industry residues) produced in Ukraine, electricity and heat generators on biomass would allow not only to obtain additional income for farmers but also the ability to guarantee the supply of biomass to numerous operators. Moreover, UABIO research has confirmed that the cultivation of 1.5% -3% of unproductive land (approximately 118 thousand hectares) creates opportunities for alternative energy production in the amount of 0.57 million tons of fuel equivalent in 5 years [10].

In addition, in 2018, the Government of Ukraine adopted the Strategy of Low Carbon Development of Ukraine until 2050. This document provides the emission reduction and the greenhouse gases absorption increase, the introduction of environmentally friendly production using "green" technologies in all sectors of the economy. Following international obligations, particularly the Paris Agreement of 2015, all member states must develop their own national strategies for low-carbon development until 2050. Ukraine was one of the first countries to develop a relevant strategic climate document and submit it to the Secretariat of the UN Framework Convention on Climate Change. The strategy identifies a possible vector of economic growth, taking into account public policy objectives to reduce greenhouse gas emissions. At the national level, the Strategy is a tool for public governance and establishing climate-responsible behaviour of business and citizens [11]. However, for effective practical implementation of this strategy, it is necessary to coordinate it with the Sustainable Development Goals of Ukraine [12] and adopt a bioeconomic development strategy to other regulatory documents of the country (Fig. 1).

The EU bioeconomic strategy [13, 14] encourages member states to develop their national bioeconomy strategies and other policies to foster links between the raw materials sectors, the bio-industries and the scientific community to achieve the strategic goals of the EU and member states [15]. Most Central and Eastern European (CEE) countries lag in developing specific national bioeconomy strategies and/or action plans that directly impact economic, environmental, and social issues [16]. One of the specific indicators is the uneven distribution of industrial and bio-processing enterprises [17], with high value-adding along the entire sustainable value chain. Developing a national bioeconomic strategy can be an excellent alternative for Ukraine, which is now dependent on fossil fuels, increasing access to sustainable and affordable renewable energy, building a "green" industrial base, preserving and ensuring the transformation of the economy in line with sustainable development goals.

the country's natural capital

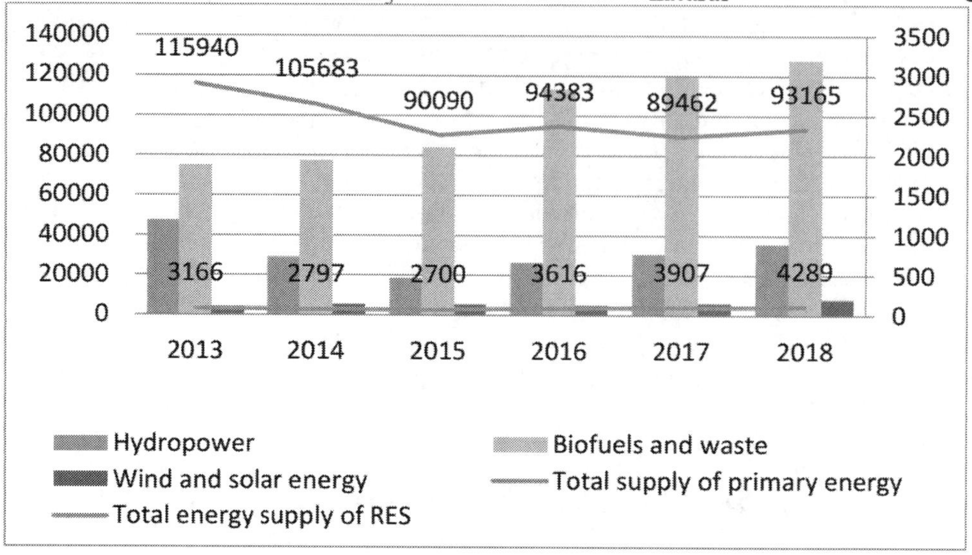

Fig. 1. Produced from alternative energy sources, thousand tons fuel equivalent.

In this regard, it is important to develop and implement a system for assessing and monitoring indicators of the impact of bioeconomic activities on achieving the Sustainable Development Goals and economic growth, improving and systematizing the regulatory framework for bioeconomy in Ukraine in accordance with Sustainable Development Goals, intersectoral balance modeling, substantiation of strategic vision and modeling of bioeconomic transformation of branches of national economy.

3 Use of biogas plants

According to NEC Ukrenergo, the power system of Ukraine already requires the use of approximately 2 GW of high-shunting capacity, which can provide at least four starts per day and have a significant range of regulation (80% of the installed capacity). The development of biogas technologies allows the efficient use of gas turbines (pickers) and medium-speed internal combustion engines (balancers) as high-shunting power. The use of biogas technologies as balancers makes it possible to reduce the forecasting error for RESs with an uneven generation schedule [18-20].

The use of biotechnology is divided into two main categories: "traditional" and "modern". "Traditional" is the use of biomass in such forms as wood, animal waste, etc. Modern bioenergy technologies include the production of liquid biofuels, the raw materials for the manufacture of which are plants; production of biogas obtained as a result of the anaerobic conversion of animal residues; heating systems on wood pellets, and others. About three-quarters of the world's renewable energy use is related to bioenergy, with more than half of it consisting of traditional biomass use. In 2015, bioenergy accounted for about 10% of total final energy consumption and 1.9% of world electricity production. Biomass already has significant potential to increase energy supply in densely populated countries such as Brazil, India, and China, and biogas can be used efficiently to increase the energy efficiency of RES with an irregular generation schedule in the power system electricity balance [21-22]. Biogas technologies can solve the most critical challenges of our time: the fight against the growing amount of organic waste generated by modern society and the economy and the need to reduce CO_2. By converting organic waste into a renewable energy resource, biogas production is one of the ways to accumulate electricity that meets the requirements of resource reuse and waste disposal and meets the growing demand for energy balancing services and environmental requirements. According to IRENA, the installed capacity of power plants that generate electricity from liquid biofuels, municipal waste, biogas, and solid biofuels grows every year (Fig. 2) [19].

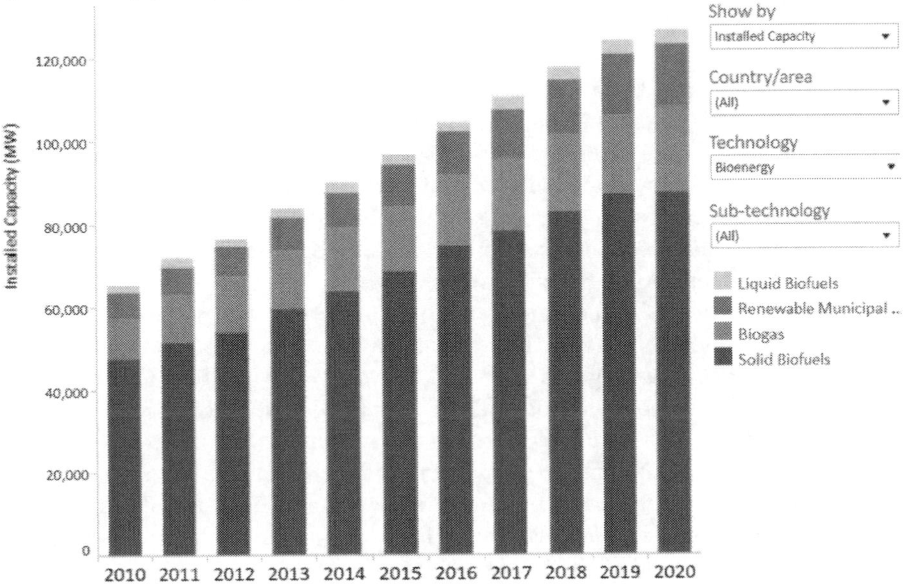

Fig. 2. Changing the installed capacity of power plants that generate electricity from liquid biofuels, municipal waste, biogas, and solid biofuels worldwide.

According to IEA forecasts, bioenergy will provide only 5% of the total amount of electricity produced in 2050, but it is already one of the sources of manoeuvrability with low CO_2 emissions.

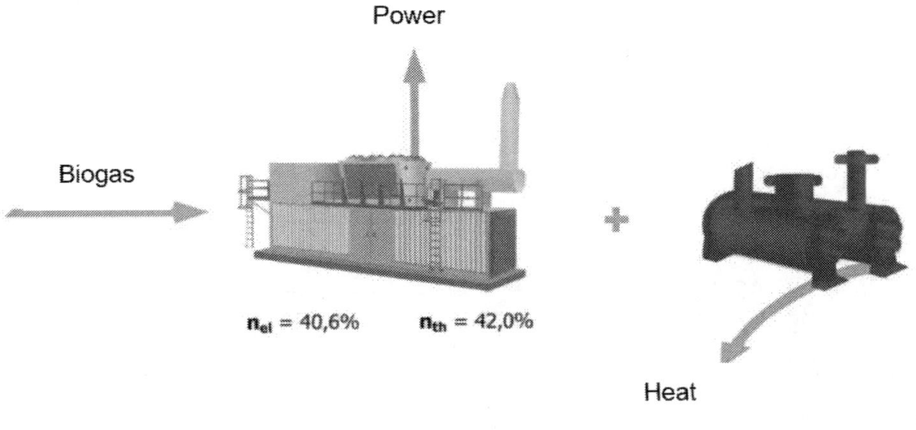

Fig. 3. Generation of electricity and heat.

Biogas is used to generate electricity, consisting of 50-75% methane, 25-45% carbon dioxide, 2-8% water vapour and impurities (O_2, N_2, NH_3, H_2, H_2S). For comparison, natural gas contains 80 to 90% of methane. The average calorific value of biogas is about 21-23.5 MJ / m3, so that one m^3 of biogas corresponds to 0.5-0.6 litres of diesel fuel or about 6 kWh. The yield of biogas from plant raw materials depends not only on the type of raw material but also on the plant's design, fermentation temperature and retention time; for example, corn silage gives about eight times more biogas per ton than cow manure [19].

4 Conclusion

The EU Bioeconomic Strategy [1, 7] encourages member states to develop their national bioeconomy strategies and other policies to foster links between the raw materials sectors, the biological industries and the scientific community to achieve the strategic goals of the EU and member states [3]. Most Central and Eastern European (CEE) countries lag in developing specific national bioeconomy strategies and/or action plans that directly impact countries' economic, environmental, and social well-being [8]. One of the specific indicators is the uneven distribution of industrial processing and bio-processing enterprises [9], which create high added value along the entire stable value chain.

Developing a national bioeconomic strategy can be an excellent alternative for Ukraine, which is now dependent on fossil fuels, increasing access to sustainable and affordable renewable energy, building a "green" industrial base, preserving the country's natural capital and transforming the economy in line with sustainable development goals.

In this regard, it is essential to develop and implement a system for evaluating and monitoring indicators of the impact of bioeconomic activities on achieving the Sustainable Development Goals and economic growth, improving and systematizing the regulatory framework for bioeconomy in Ukraine following Sustainable Development Goals, intersectoral balance modelling, substantiation of strategic vision and modelling of bioeconomic transformation of branches of the national economy.

References

1 V. Vostryakova, *Bioeconomic Potential of Agrologistics Losses in Sustainable Value Chains.* Economy. Finances. Management: current issues of science and practice, **4**, (2018).

2 State Statistics Service. URL: http://www.ukrstat.gov.ua/

3 National Institute for Strategic Studies. Results of Ukraine's foreign trade for the first half of 2020 URL: https://niss.gov.ua/doslidzhennya/ekonomika/pidsumki-zovnishnoi-torgivli-ukraini-za-pershe-pivrichchya-2020-r

4 Central Observatory of Ukraine. URL: http://cgo-sreznevskyi.kyiv.ua/

5 Ministry of Environmental Protection and Natural Resources of Ukraine. URL: https://mepr.gov.ua/

6 Ministry of Agrarian Policy and Food of Ukraine. Exports of cereals, legumes and flour from Ukraine. URL: https://minagro.gov.ua/en/investoram/monitoring-stanu-apk/eksport-z-ukrayini-zernovih-zernobobovih-ta-boroshna

7 European Commission. Strengthening the link between the economy, society and the environment; European Commission (EC): Brussels, Belgium (2018).

8 European Commission. Review of the European Bioeconomy Strategy 2012; European Commission (EC): Brussels, Belgium (2017).

9 MEA. Energy balances of non-OECD countries, OECD / IEA, Paris (2015).

10 Bioenergy Association of Ukraine (UABIO). Prospects for the development of bioenergy as a tool for replacing natural gas in Ukraine, UABIO, Kyiv (2015).

11 European integration portal. Ukraine is in the top ten in the implementation of the Paris Agreement on Climate Change. URL: https://eu-ua.kmu.gov.ua/novyny/ukrayina-v-desyatci-lideriv-z-realizaciyi-paryzkoyi-ugody-shchodo-zminy-klimatu-minpryrody

12 On the Goals of Sustainable Development of Ukraine for the period up to 2030: Decree of the President of Ukraine of September 30, 2019 № 722. URL: https://zakon.rada.gov.ua/laws/show/722/2019#Text

13 COM. Innovation for Sustainable Growth: A Bioeconomy for Europe, EC (2012).

14 COM. Sustainable bioeconomy for Europe - Strengthening the link between the economy, society and the environment. Updated bioeconomy strategy (2018).

15 Panel of countries with bioeconomics from the Center for Knowledge of Bioeconomics of the European Commission. URL: https://knowledge4policy.ec.europa.eu/visualisation/bioeconomy-different-countries_en

16 Biological industry and bioprocessing plants in the EU. URL: https://knowledge4policy.ec.europa.eu/visualisation/bio-based-industrybiorefineries-eu_en

17 Chemical and material bioprocessing in the EU. URL: https://knowledge4policy.ec.europa.eu/visualisation/chemical-materialbiorefineries-eu_en

18 V. Vostriakova, The Renewable Energy Sources Contribution to Sustainable Economic Growth, in *2021 IEEE 2nd KhPI Week on Advanced Technology (KhPIWeek)*, 2021, 381-386, doi: 10.1109/KhPIWeek53812.2021.9569993

19 The International Renewable Energy Agency (IRENA). https://www.irena.org/

20 G. S. Lakshmi, O. Rubanenko, and I. Hunko, *Control of the Sectioned Electrical Network Modes with Renewable Energy Sources,* in Proceedings of the International Conference on Sustainable Energy and Future Electric Transportation, SeFet (2021).

21 G. S. Lakshmi, O. Rubanenko, G. Divya, and V. Lavanya, *Distribution Energy Generation using Renewable Energy Sources*, in Proceedings IEEE India Council International Subsections Conference, INDISCON, 2020, 108-113, India (2020).

Assessing the Involvement of Industries in Global Value Chains

Viktoriia Khaustova[1], *Olena* Reshetnyak[1*], *Mykyta* Khaustov[1], *Taras* Danko[2], and *Natalia* Danko[3]

[1]National Academy of Sciences of Ukraine, Research Centre of Industrial Problems of Development, 1a Inzhenernyi Ln., Kharkiv, 61166, Ukraine
[2]National Technical University «Kharkiv Polytechnic Institute», 2 Kyrpychova str., 61002, Kharkiv, Ukraine
[3]V. N. Karazin Kharkiv National University, 4 Svobody Sq., Kharkiv, 61022, Ukrainian

Abstract. The purpose of the article is to assess the involvement of industries in global value chains and determine the prospects for intensification of these processes. The research methods are statistical analysis, comparison, graphical analysis. The role of export volumes of high value-added products, their enclosing in global value chains is determined. A methodical approach to assessing the involvement of industries in global value chains is proposed, allowing to detect the state of the country's involvement in the international distribution of labour in concordance with specific industries. The analysis of the structure of exports of industries on the example of Ukraine is carried out, the industries that have a significant export share both in the total exports and in the structure of production output are determined. The leading indicators of the country's involvement in global value chains are computed, as follows: the national added value in the country's exports by industries, the contribution of specific industry sectors of the economy to the national added value of gross exports, the share of the national added value in exports of industries in the total national added value of gross exports. The positioning of industries within the coordinate system «share in the export structure – share of the national added value in exports» is carried out. It is determined that the share of national added value in exports of extractive industries of Ukraine is higher compared to the average of the world countries in this research. At the same time, being lower in the industries of the manufacturing industry, thus justifying the virtual absence of knowledge-intensive and innovative stages of production of the manufacturing industry, which necessitates a deeper specialization in the exports of low-grade products. The directions of a rational integration of the Ukrainian economy into global value chains are substantiated. The measures to be assumed on the part of the government policy to intensify the involvement of Ukrainian producers in global value chains are proposed. Keywords: national value-added in exports, value chain, industry

* Corresponding author: olena.reshetnyak@karazin.ua

1 Introduction

Under the current conditions, in the profound development of globalization processes, the role of foreign trade increases due to the consideration of the national economy as not a separate economic system but a system complex of interrelations of economic entities in the worldwide market [1]. The latest trends in the development of world trade are characterized by the increasing influence of internationalization and fragmentation of production on the growth of interdependence of countries within global value chains [2]. This causes the long-term socio-economic development of the country to be conditional on the integration of the country's industries into the global economy, the volumes of exports of products with high added value, and their enclosure in global value chains. So, the country's winnings from foreign trade are determined by its involvement in the vertical specialization of the worldwide market and ensuring the creation of added value. The low level of the national added value in exports of specific industries characterizes the high level of dependence of the industry on import supplies. Despite the growing demand from economists and practitioners, an assessment of the value-added in exports cannot be implemented within the framework of traditional statistics of foreign economic activity, which records gross export and import flows in both the cost and the physical volumes between the national economies. The demand for statistical information concerning international trade-in value-added caused the creation in 2013 of an appropriate database (TiVA) by the Organization for Economic Cooperation and Development (OECD) in cooperation with the World Trade Organization (WTO). The TiVA database contains 46 indicators for 64 countries (Ukraine is omitted therein) and for 34 sectors of industry and services.

The use of value-added trade information allows to increase the degree of substantiation of the country's economic policy: to increase the accuracy in the assessment of the openness of the economy, the impact of the national currency on the foreign trade flows, to determine the sources of growth of competitiveness of the national products in foreign markets.

The relevance of the problematics of value-added trade assessment is justified in the works of many authors. In one of the first researches devoted to the issues of value-added trade, i. e., the study by D. Hummels, J. Ishii, and K. Yi [3], the authors substantiated the need to determine the foreign component in exports. The approach suggested in that study was further developed in R. Coopman and Z. Wang [4], R. Jonson [5], X. Chen [6], which, in turn, based on Leontiev's matrix of «costs-output», had proposed their methods of studying the national added value in exports, including the value created in different sectors of the economy. A unique UNCTAD study [7], published in 2013, identified the relationship between enclosing in value chains and the economic development of countries worldwide. The assessment of the national value-added impact in exports on the economic development of the world countries was analyzed in the publications by M. Kyzym [8, 9], I. Gryshova [9, 10], the countries not included in the TiVA database (e. g., Belarus) were considered in the works by V. Krivorotov [11], A. Bykov [12, 13], I. Tochitskaya [14], and others. However, because of the results of literary sources analysis, most of the indicators offered by the TiVA database do not concern the economy of Ukraine; hence an assessment of enclosing the individual industries of Ukraine in global value chains along with the determination of the state of involvement of the country in the international division of labour appears to be relevant.

2 Materials and Methods

The article aims to assess the involvement of Ukrainian industries in global value chains and determine the prospects for intensification of these processes.

The object of research is the assessment of the involvement of Ukrainian industries in global value chains.

The methodical approach to assessing the involvement of Ukrainian industries in global value chains includes the following stages:

1. Analysis of the structure of exports of Ukrainian industries and determination of industries that have a significant share of exports in total exports and production output structure.

2. Assessment of the share of national added value in exports by industries of Ukraine. Based on the OECD value-added trading indicators (TiVA) [15], the authors compute the leading indicators of the country's involvement in global value chains (the study does not identify individual partner countries but analyzes the aggregate exports by the industry sectors or agriculture):

– the national added value in the country's exports by industry and industry sector, which characterizes the added value created by the export industry sector in the production process, as well as any added value received from the domestic suppliers and embodied in the exports:

$$EXGR\_DVA_i = V * B * EXGR_i \qquad (1)$$

where **V** – vector of dimension **K**×1, where each element is equal to the specific weight of the gross added value of the industries of the country;

B – global matrix of complete material costs corresponding to the country without taking into account the costs of imports: $B = (I - A)^{-1}$, **A** – global matrix of direct material costs;

$EXGR_i$ – gross export of the *i*-th vector of dimension **K**×1, in which all elements are zero, except that corresponds to the *i*-th industry;

– contribution of the *i*-th industry to the national added value of gross exports. The indicator reflects the share in the total value of gross exports of the national added value of exports of the *i*-th industry:

$$EXGR\_DVASH_i = \frac{EXGR\_DVA_i}{EXGR_i} \times 100 \qquad (2)$$

– share of the national added value of exports of the *i*-th industry in the total national added value of gross exports:

$$EXGR\_TDVIND_i = \frac{EXGR\_DVA_i}{\sum_i^N EXGR\_DVA_i} \times 100 \qquad (3)$$

3. Positioning industries in the matrix's quadrants in the system of coordinates «share in the export structure – share of the national added value in exports». As a result, industries with a share in the export structure and a share of the national added value in exports above the average level in the country's economy are determined. For these industries, the share of national added value in exports is compared with other world countries.

4. Substantiation of a rational integration of Ukraine's economy into global value chains.

3 Results

Following the proposed methodical approach on the grounds of the analysis of data of the matrix of «costs-output» of Ukraine based on the data of 2018, the rating of industries is carried out (for comparison, in the analysis are included the agricultural, fisheries and forestry industry sectors) by the indicator of the industry share in the export structure as presented by the Fig. 1.

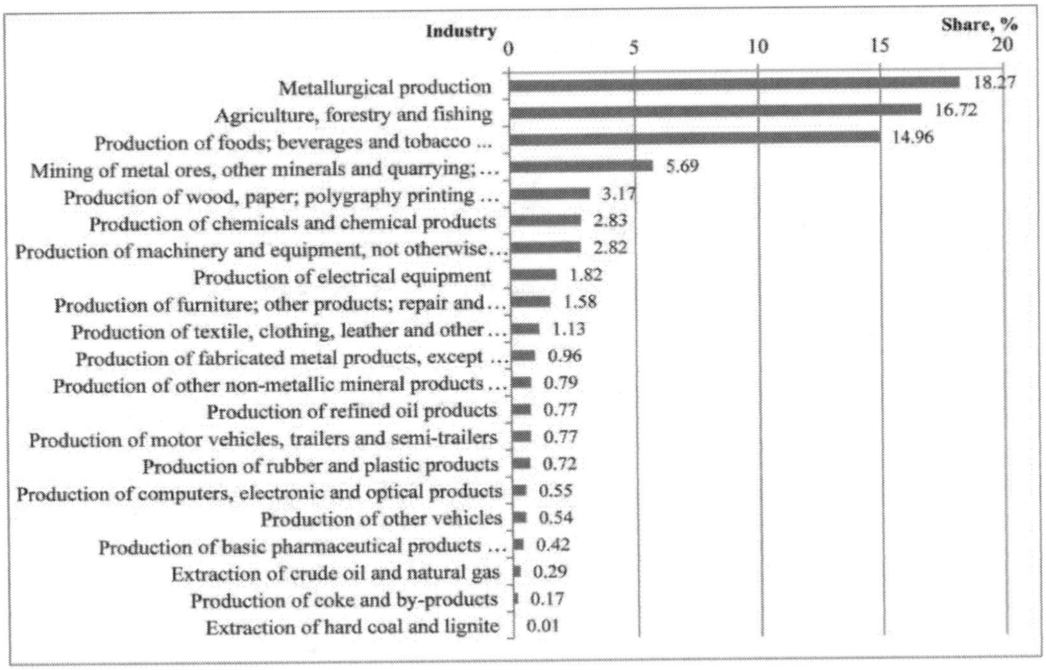

Fig. 1. The rating of industries by share in the export structure according to the year 2018 Source: computed by the authors based on the data [16].

According to the rating, the largest share in the structure of exports of Ukraine's industry in 2018 is held by metallurgical production (18.27%), in the second place are agriculture, forestry and fisheries (16.72%), in the third-place – production of food and beverages (14.96%). It should be noted that high-tech industries – production of computers, electronic and optical products, and essential pharmaceutical products and pharmaceutical preparations have a small share in the export structure (respectively, 0.55% and 0.45%).

The study of the share of exports in the production of the individual industry sectors, agriculture, forestry and fisheries has shown that the most significant share is held by the exports in metallurgical production (63.57%), the production of electrical equipment holds the second place (60.45%), the production of machinery and equipment – the third place (59.65%), according to the data of the year 2018. The rating of industry sectors, agriculture, forestry and fisheries is presented in Fig. 2.

Also, the indicator of the share of national added value in exports of individual industries, agriculture, forestry and fisheries is computed, characterizing the contribution of specific sectors of the economy to global value chains. The rating of industries by this indicator is presented in Fig. 3.

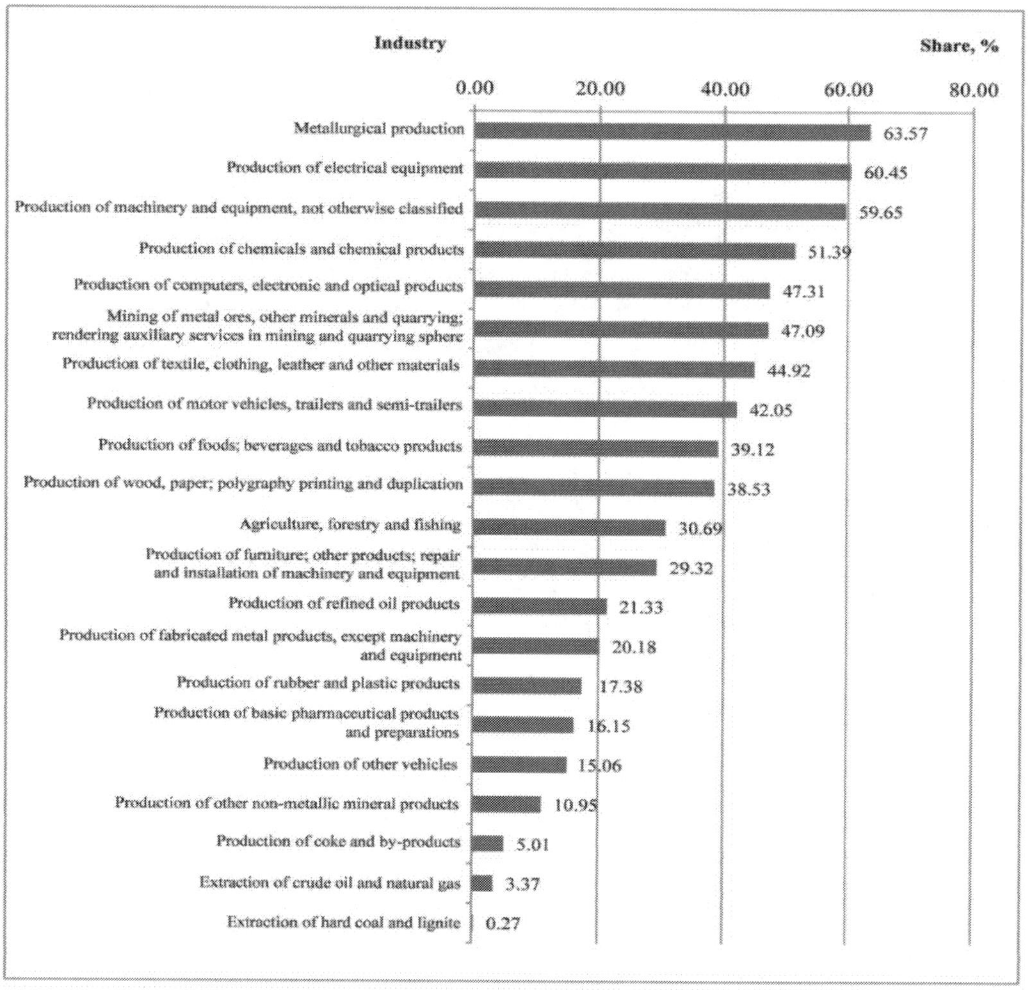

Fig. 2. The rating of individual industries in terms of export share in the production output of individual industries Source: computed by the authors based on the data [16].

Thus, the share of national added value in exports above average (62.21%) is shown by the following industries: extraction of hard coal and lignite (85.57%), extraction of crude oil and natural gas (82.91%), production of other non-metallic mineral products (79.46%), production of foods; beverages and tobacco products (76.99), metallurgical production (76.77%), mining of metal ores, other minerals and quarrying; rendering auxiliary services in mining and quarrying sphere (75.54%), agriculture, forestry and fishing (71.01%), production of fabricated metal products, except machinery and equipment (70.26%), production of wood, paper; polygraphy printing and duplication (65.59%), production of essential pharmaceutical products and preparations (65.53%), production of furniture; production of other products; repair and installation of machinery and equipment (64.62%) and production of other vehicles (63.84%). Other industries have a high level of dependence on import supplies, which exceeds the average level of the total national added value in exports by individual industries of Ukraine. Other industries have a high level of dependence on import supplies, which exceeds the average level of the total national added value in exports by individual industries of Ukraine.

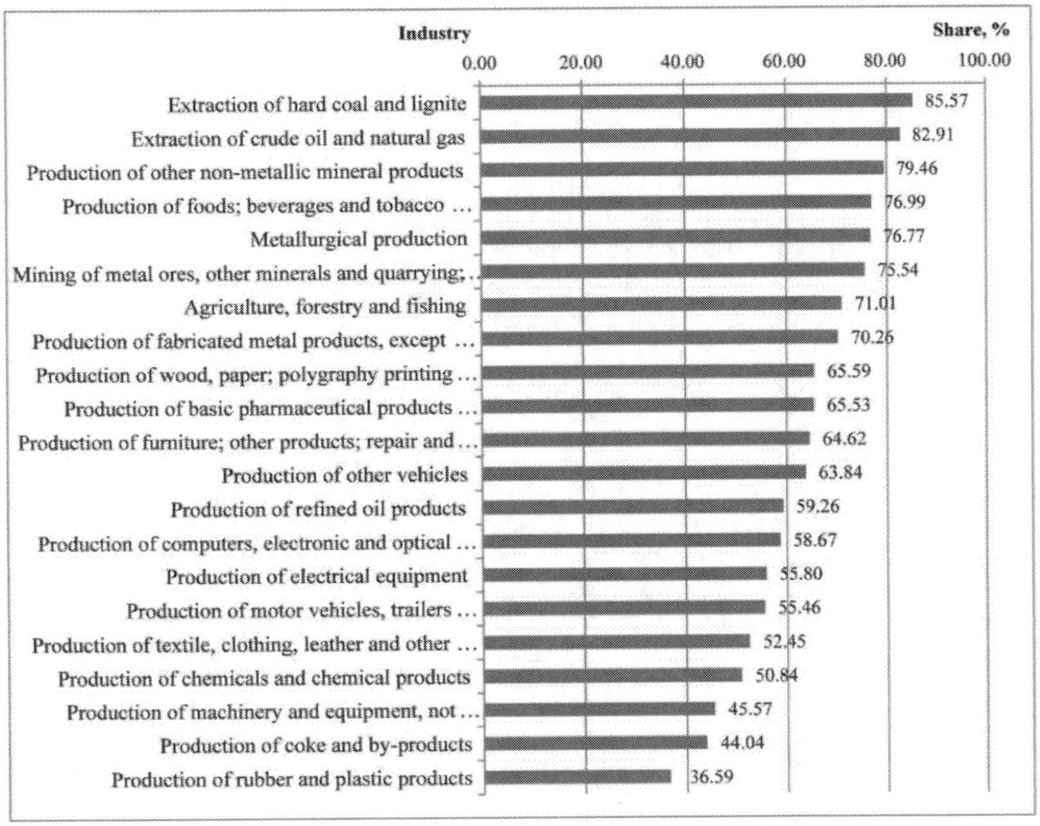

Fig. 3. The rating of individual industries by share of the national added value in their exports
Source: computed by the authors based on the data [16].

It is also purposeful to promote the scientific support for the growth of industries with a more significant share in the export structure and a share in the structure of national added export value compared to other sectors of the economy. The computation of the Pearson correlation coefficient between the export structure and the share of national added value in exports of industries has shown a lack of correlation (–0.03), allowing to position industries in the matrix quadrants in the system of coordinates «share in the export structure – share of national added value in exports».

The matrix of the positioning of industries in the system of coordinates «share in the export structure – share of national added value in exports» in 2018 consists of four quadrants: I – «share in the export structure is below 1.22% – share of national added value in exports is below 62.21%»; II – «share in the export structure is above 1.22% – share of national added value in exports is below 62.21%»; III – «share in the export structure is below 1.22% – share of national added value in exports is above 62.21%»; IV – «share in the export structure is above 1.22% – share of national added value in exports is above 62.21%». Positioning of Ukrainian industries in the quadrants of the matrix in the system of coordinates «share in the export structure – share of national added value in exports» in 2018 is presented in Table 1.

Based on the carried out positioning of industries in the quadrant IV, which characterizes the shares in the structure of exports and national added value in exports above the average level, the following industries are included – production of foods; beverages and tobacco products; metallurgical production; mining of metal ores, other minerals and quarrying; rendering auxiliary services in mining and quarrying sphere;

agriculture, forestry and fishing; production of wood, paper; polygraphy printing and duplication.

Table 1. The matrix of the positioning of Ukrainian industries in the quadrants of the system of coordinates «share in the export structure – share of national added value in exports» in 2018

Indicator		Share of the national added value in exports, %	
		Below average 1.22%	Above-average 1.22%
Share of the national added value in exports, %	Above average (62.21%)	Extraction of hard coal and lignite Extraction of crude oil and natural gas; Production of other non-metallic mineral products Production of fabricated metal products, except machinery and equipment Production of essential pharmaceutical products and preparations Production of furniture; other products; repair and installation of machinery and equipment Production of other vehicles	Production of foods; beverages, and tobacco products Metallurgical production Mining of metal ores, other minerals and quarrying; rendering auxiliary services in mining and quarrying Agriculture, forestry and fishing Production of wood, paper; polygraphy printing and duplication
	Below average (62.21%)	Production of textile, clothing, leather and other materials Production of computers, electronic and optical products Production of electrical equipment Production of motor vehicles, trailers and semi-trailers Production of refined oil products Production of rubber and plastic products Production of coke and by-products	Production of machinery and equipment, not otherwise classified Production of chemicals and chemical products

Source: computed by the authors based on the data [16].

Therefore, to ensure the development of the country's economy, it is necessary to pay more attention to those industries that provide a high share in the export structure and have a high share of the national added value in exports.

The next step of the analysis is to compare the national added value in exports of the industries that ensure the growth of exports and national added value in exports with other world countries.

To compare the share of national added value in Ukraine's exports with similar indicators of the world countries, the value-added trading database (TiVA) [17] is used. According to the indicator of national added value in exports (the rating of the world countries is presented by the Fig. 4), Ukraine ranks 19th (83.6%), which is higher than the average level (73.4%).

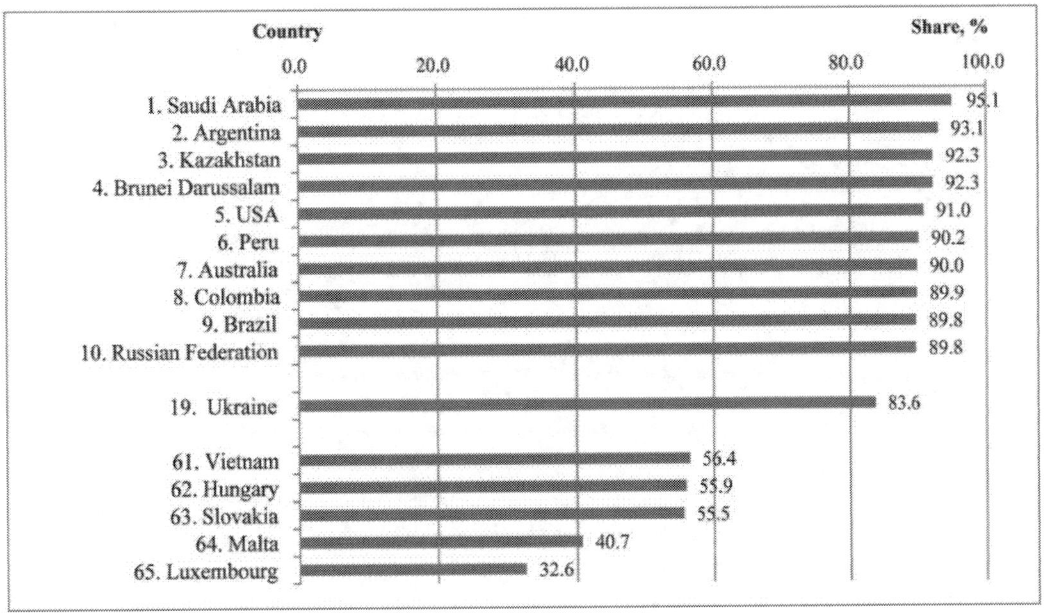

Fig. 4. The share of national added value in exports of the world countries according to the data of the year 2018. Source: computed by the authors based on the data [16, 17].

Studies of the share of national added value in exports of the individual industries of the economy have shown that in the industry of agriculture, forestry, and fisheries, Ukraine ranks 38th among 65 countries of the world (80.8%), less than the average level of 88.3% (Fig. 5).

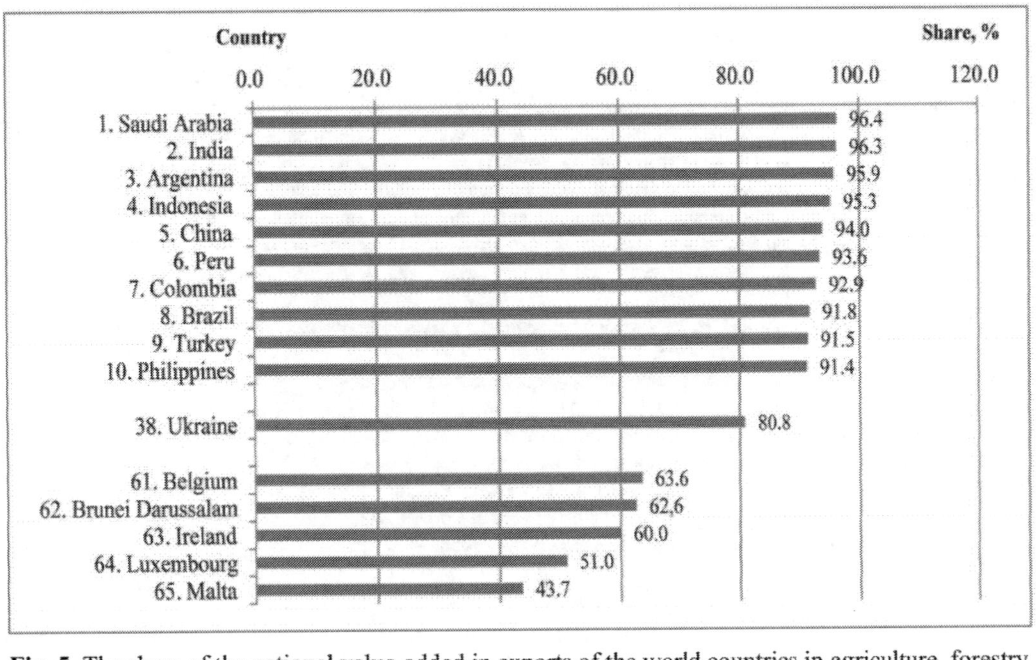

Fig. 5. The share of the national value-added in exports of the world countries in agriculture, forestry and fisheries in 2018. Source: computed by the authors based on the data [16, 17].

In mining non-energy products, Ukraine ranks 20th out of 65 countries (86.5%), i.e., above the average level of 73.4% (Fig. 6).

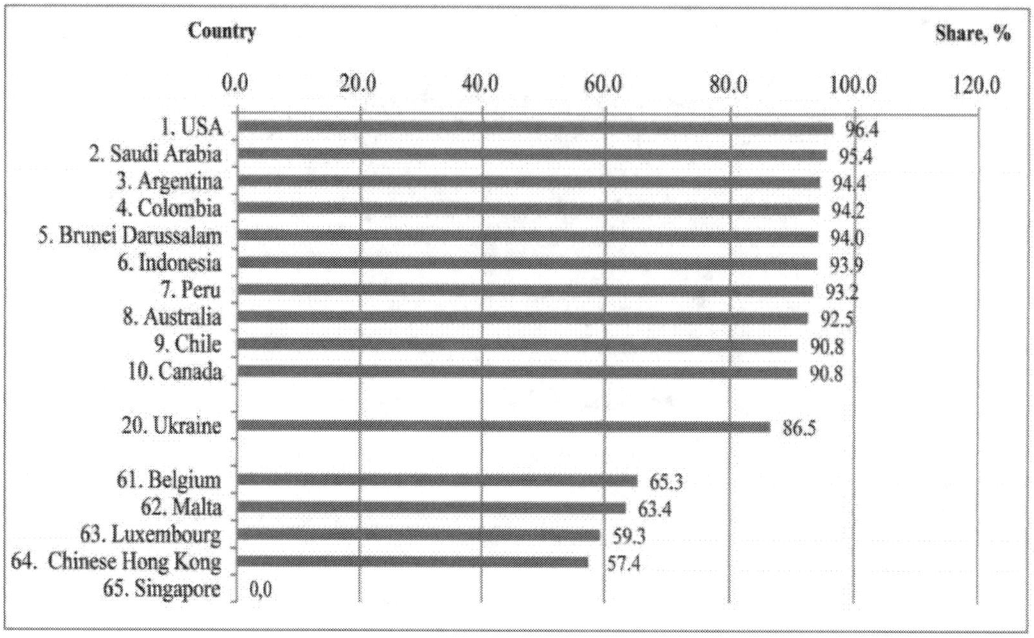

Fig. 6. The share of national added value in exports of the world countries in mining non-energy products in 2018. Source: computed by the authors based on the data [16, 17].

In the industry of production of foods, beverages and tobacco, Ukraine ranks 22nd out of 65 countries (81.5%), which is also above the average level of 76.9% (Fig. 7).

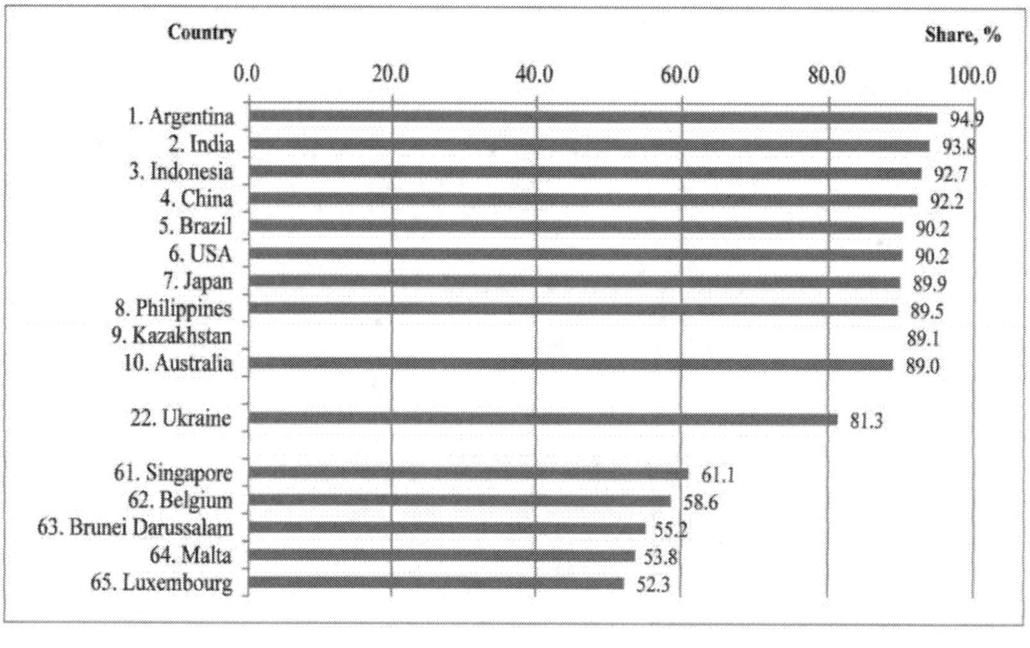

Fig. 7. The share of national added value in exports of the world countries in the production of foods, beverages and tobacco in 2018. Source: computed by the authors based on the data [16, 17].

In the production of chemical and pharmaceutical products, Ukraine ranks 45th out of 65 countries (65.2%), below the average level of 77.6% (Fig. 8).

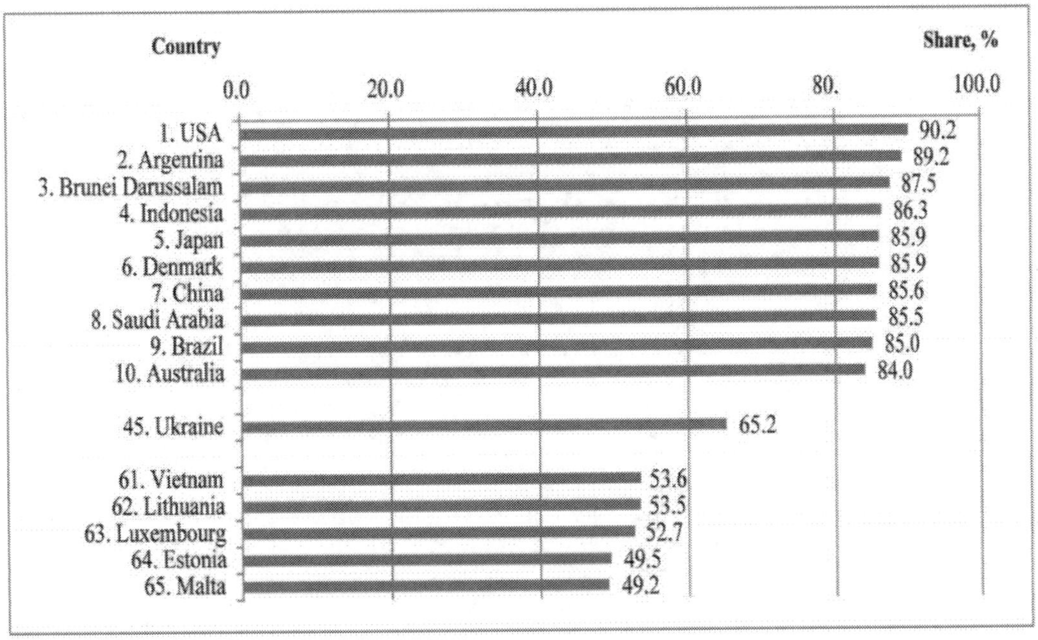

Fig. 8. The share of national added value in exports of the world countries in chemical and pharmaceutical products in 2018. Source: computed by the authors based on the data [16, 17]

In the industry of metallurgical production, Ukraine ranks 45th out of 65 countries (65.2%), i. e., below the average level of 77.6% (Fig. 9).

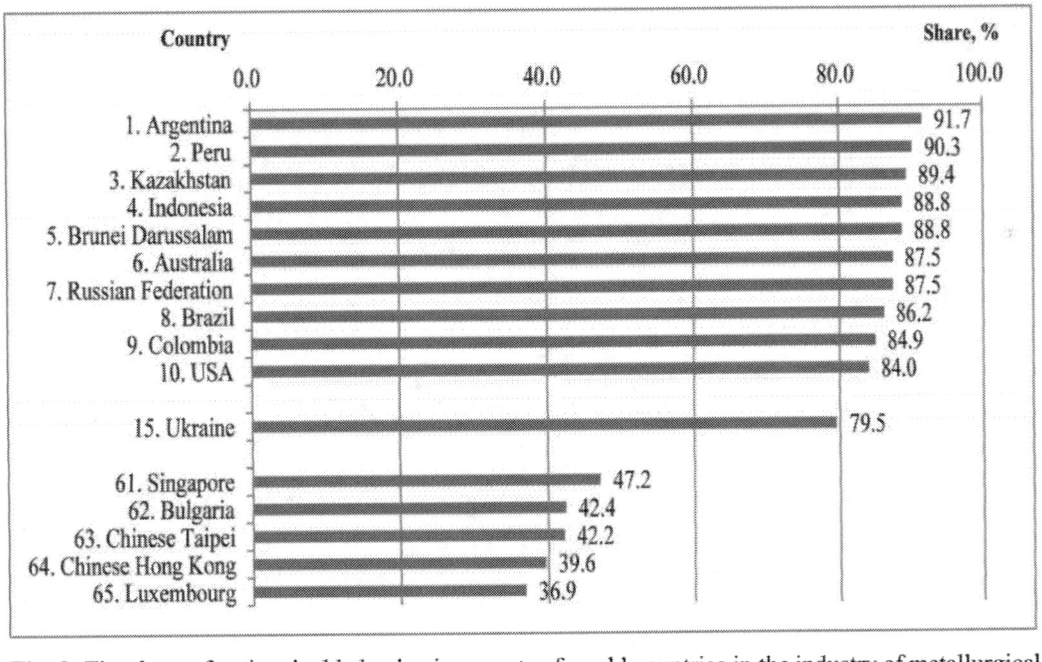

Fig. 9. The share of national added value in exports of world countries in the industry of metallurgical production in 2018. Source: computed by the authors based on the data [16, 17].

In the computer, electronic and optical production industry, Ukraine ranks 34th out of 65 countries (70.2%), which is below the average level of 71.7% (Fig. 10).

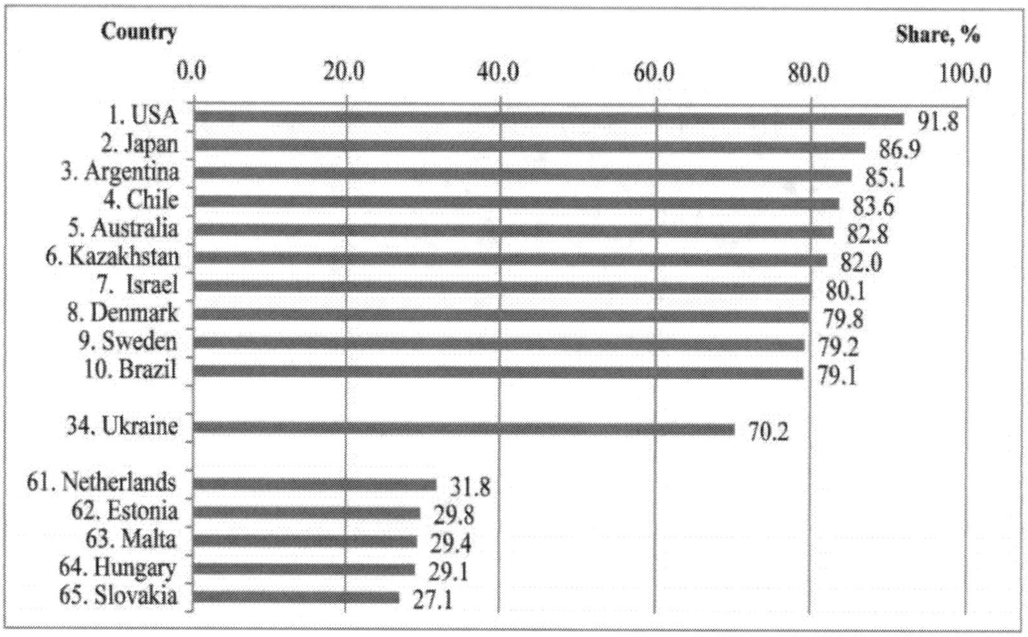

Fig.10. The share of national added value in exports of world countries in computer, electronic and optical production in 2018. Source: computed by the authors based on the data [16, 17].

In the industry of electrical equipment production, Ukraine ranks 33rd out of 65 countries (64.6%), which is below the average level of 66.8% (Fig. 11).

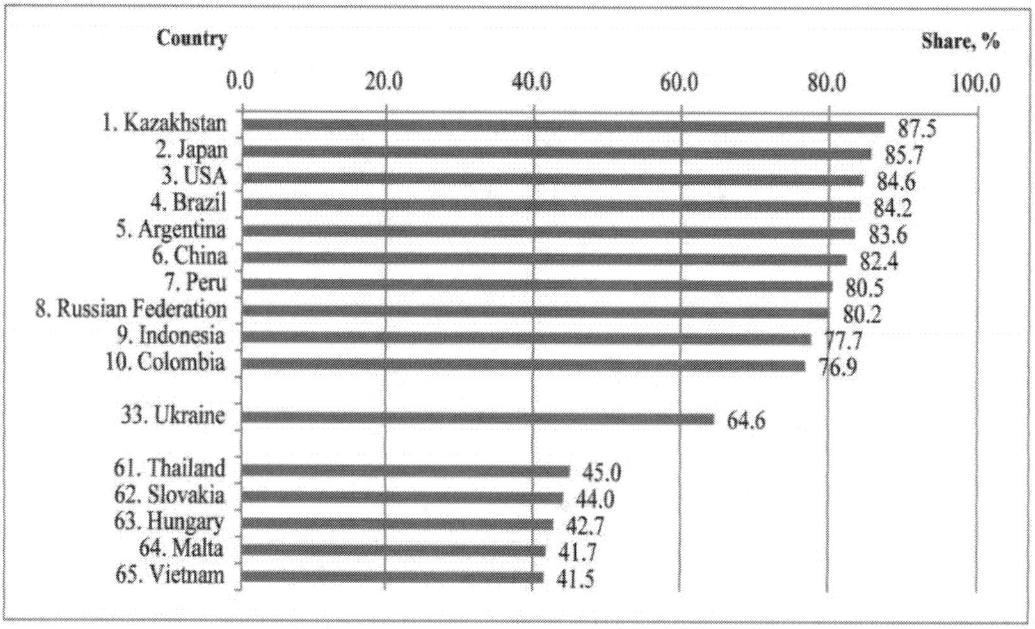

Fig. 11. The share of national added value in exports of world countries in the production industry of electrical equipment in 2018. Source: computed by the authors based on the data [16, 17].

In the machinery and equipment production industry, Ukraine ranks 49th out of 65 countries (61.8%), below the average level of 76.9% (Fig. 12).

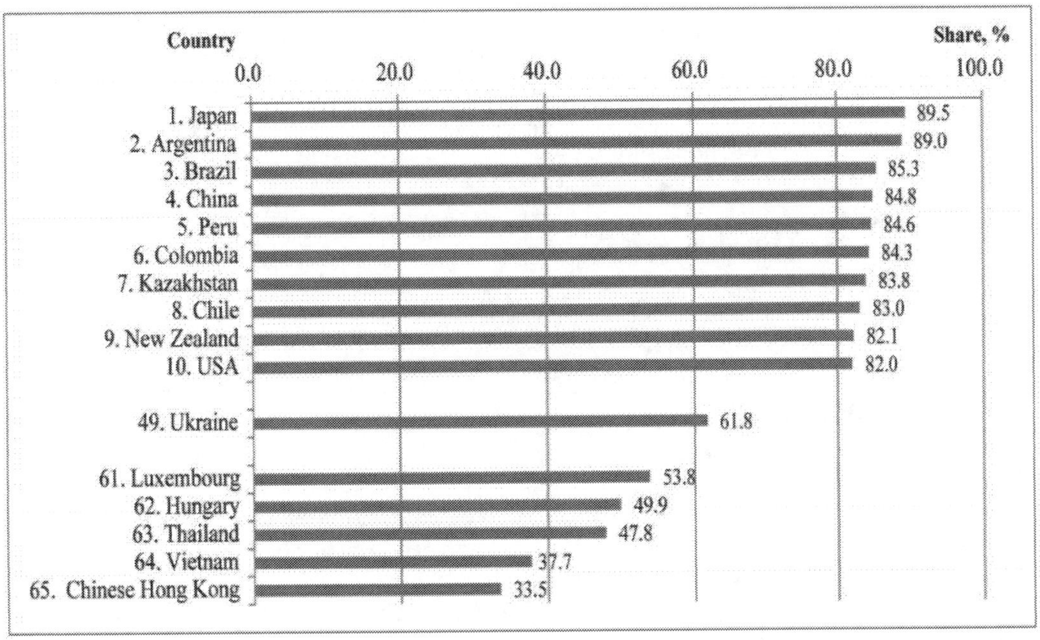

Fig. 12. The share of national added value in exports of world countries in the industry of machinery and equipment production in 2018. Source: computed by the authors based on the data [16, 17].

Ergo, the carried out analysis of the share of national added value in the exports of Ukraine allows us to conclude that this indicator is generally higher for the mining industries than other world countries. In comparison, this indicator is lower for the manufacturing industries, which justifies the domestic industry's dependence on imported components based on foreign technologies. There are virtually no knowledge-intensive and innovative stages of production of machine-building products in Ukraine. Technological lag and reduction of industrial production, which has been observed for a long time in the country's economy, have caused the decline of national high-tech industries [18]. This state and tendencies in the development of domestic industries have led to the deepening of Ukraine's specialization in the exports of products with low-level processing.

The need to reduce the impact of risks in terms of foreign economic security and attract foreign investments for the industrialization of the Ukrainian economy can be achieved by increasing the concentration of knowledge-intensive and innovative stages of production in the industry.

Given the world experience, the main directions of the State policy on intensification of the involvement of Ukrainian producers in global value chains can be formulated as follows:

– need's identification to increase the concentration of knowledge-intensive and innovative stages of production in the industry as crucial support for economic development at the level of the economic strategy of the State;

– use of development institutions to ensure the necessary conditions for the integration of Ukrainian enterprises into global value chains (national financial development institutions have the most significant potential of the State support in the studied aspect);

– the State support for exports, which, in addition to the State-based programming and institutional provision, should also include improvement of the regulatory environment; formation of innovative market infrastructure with the transfer of innovations; accumulation of funds for the scientific researches necessary for exporters; support of the financial leasing and franchising; training personnel in the field of international trade; reducing the tax pressure; ensuring the partnership between local authorities and entrepreneurial associations; participation of local authorities in the regional and interregional trade associations; spread of information about the export and investment opportunities of enterprises in the region; provision of consulting services, etc.;

– special economic regimes of a territorial nature, which causes the use of the compact location of high-tech productions in the context of the inclusion of their products in global value chains;

– subsystem of the tax policy instruments, including special economic zones and tax regimes with the industry orientation, the local (elemental) tax benefits.

The defined vital directions of the State support are interrelated and interdependent, should be used simultaneously and are being considered as subsystems that comprise a system of State support for involving Ukrainian producers in global value chains.

4 Conclusions and Discussion

Thus, as a result of the carried out study, we can specify the following:

1. In modern conditions of the deepened globalization processes, the role of determination of the volume of exports of products with high added value, as well as their enclosing in global value chains, is increasing
2. The methodical approach to assessing the enclosing of Ukrainian industries in global value chains is proposed, which allows determination of the state of involvement of the country in the international division of labour.
3. The most significant shares in the export structure of Ukraine's industry in 2018 belong to metallurgical production (18.27%), agriculture, forestry and fisheries (16.72%), food and beverage production (14.96%). At the same time, high-tech industries in Ukraine have a small share in the export structure.
4. The study of the share of exports in the production of individual industries has shown that the most significant shares of exports in production output belong to metallurgical production (63.57%), production of electrical equipment (60.45%), machinery and equipment production (59.65%).
5. The following industries demonstrate a high share of the national added value in exports: black and brown coal mining (85.57%), crude oil and natural gas production (82.91%), production of foods; beverages and tobacco products (76.99%), metallurgical production (76.77%), metal ore mining, other minerals and quarrying (75.54%), agriculture, forestry and fisheries (71.01%) – mining industry mainly.
6. The carried out positioning of industries in the system of coordinates «share in the export structure – share of national added value in exports» allowed to allocate the industry sectors of the economy that have shares in the structure of exports and national added value of exports above the average level, with a high impact on the economic activity of Ukraine as an exporting country.
7. The study of the share of national added value in Ukrainian exports compared to other world countries has shown that in the mining industry, this indicator is higher. In contrast, in the manufacturing industries, the average level is lower, justifying the dependence of the domestic industry on imported components based on foreign technologies. In Ukraine, there are virtually no knowledge-intensive and innovative stages of production of machine-building products, which has led to the deepening of Ukraine's specialization in the export of low-processed products.
8. The directions of the State policy on intensification of involvement of Ukrainian producers in global value chains are proposed and should function simultaneously for their more significant effect, including identification of the need to increase the concentration of knowledge-intensive and innovative stages of production in the industry at the level of the economic strategy of the State; use of development institutions; the State support for

exports; special economic regimes of a territorial nature; subsystem of the tax policy instruments.

References

1. V. Khaustova, O. Reshetnyak, R. Lobodin, *Evaluating the Impact of the Scientific Activity of a Country on the Provision of its Economic Security,* 36th IBIMA Conference: 4-5 November, Granada, Spain (2020). URL: https://ibima.org/accepted-paper/evaluating-the-impact-of-the-scientific-activity-of-a-country-on-the-provision-of-its-economic-securi
2. M. Kyzym, O. Reshetnyak, O. Kozyrieva, V. Khaustova, *Research Organization on The Basis of The World Foresights Analysis: Trends and Scenario*s, 35th IBIMA Conference: 1-2 April, Seville, Spain. (2020). URL: https://ibima.org/accepted-paper/research-organization-on-the-basis-of-the-world-foresights-analysis-trends-and-scenarios
3. D. Hummels, J. Ishii, K-M. Yi, Journal of International Economics, **54**, Iss. 1, 75–96 (2001) DOI: https://doi.org/10.1016/S0022-1996(00)00093-3
4. R. Koopman, Z. Wang, Sh-J Wei, *How Much of Chinese Exports Is Really Made in China? Assessing Foreign and Domestic Value-Added in Gross Exports,* Office of Economics Working Paper, U. S. International Trade Commission, 2008-03-B. (2008) DOI: 10.3386/w14109.
5. R. Jonson. Five, Journal of Economic Perspectives, **28**, 119–142 (2014) DOI: 10.1257/jep.28.2.119.
6. X. Chen., et al. *The Estimation of Domestic Value-Added and Employment Induced by Exports: An Application to Chinese Exports to the United States. In China and Asia: Economic and financial interactions*. Oxon: Routledge, 64–82 (2009)
7. Global Value Chains and Development: Investment and Value Added Trade in the Global Economy: A preliminary analysis. UNCTAD, DIAE, 40 (2013) URL: https://unctad.org/system/files/official-document/diae2013d1_en.pdf
8. M. Kyzym, V. Khaustova, V. Shlykova, H. Kramarev, A. Pinchuk, The Problems of Economy, **4**, 19–29 (2018) DOI: https://doi.org/10.32983/2222-0712-2018-4-19-29
9. I.Gryshova, M. Kyzym, V. Khaustova, V. Korneev, H.Kramarev, Sustainability, **12**, 2072 (2020)DOI:10.3390/su12052072
10. I. Gryshova, M. Kyzym, I. Hubarieva, V. Khaustova, A. Livinskyi, M. Koroshenko, Sustainability, **12**, 7692 (2020) DOI:10.3390/su12187692
11. V. Krivorotov, V. Mokhov, O. Ivanova, O. Polyakova, *Convergence modelling in international integration associations*, Bulletin of the South Ural State University. Ser. Mathematical Modelling, Programming & Computer Software (Bulletin SUSU MMCS). **13**, 68–72 (2020) DOI: 10.14529/mmp200306
12. A. Bykov, O. Kolb, Belarusian Economic Journal, **1**, 128–143 (2016) URL: http://edoc.bseu.by:8080/bitstream/edoc/30257/1/Bykov_A._A.._s._128_143.pdf
13. A. Bykov et al., Scientific works of the Belarusian State Economic University, **12**, 63–72 (2019) URL: https://ndc-ipr.org/media/publications/files/Bykov_A.A._s_63_72.pdf
14. I. Tochitskaya, *Assessment of value added in the export of Belarus,* Bankaўski Vesnik. KRASAVIK, 3–11 (2020) URL: http://www.nbrb.by/bv/articles/10746.pdf
15. Guide to OECD's Trade in Value Added (TiVA) Indicators. OECD (2019) URL: https://www.oecd.org/sti/ind/tiva/TiVA2018_Indicators_Guide.pdf
16. Tables "input-output" (in consumer prices) for 2013-2018 / Official site of the State Statistics Service of Ukraine. URL: http://ukrstat.gov.ua/operativ/operativ2006/vvp/vitr_vip/vitr_u/arh_vitr_u.html

17. Trade in Value Added (TiVA): Principal indicators. OECD. Stat. URL: https://stats.oecd.org/Index.aspx?DataSetCode=TIVA_2018_C1
18. O. Reshetnyak, Economy and State, **9**, 44–51 (2020) DOI: 10.32702/2306-6806.2020.9.44

Information and Communication Technologies Usage in Working with Families of Early Aged and Preschool Children with Visual Impairments in Terms of Distance Psychological and Pedagogical Support

Tatiana Kostenko[1*], and *Kateryna* Dovhopola[1]

[1]Department of Education of Visually Impaired Children, Institute of Special Pedagogy and Psychology named after Mykola Yarmachenko, National Academy of Pedagogical Sciences of Ukraine, 9, M. Berlinskoho St., Kyiv, 04060, Ukraine Kyiv, Ukraine

Abstract. This paper presents the results of studying the use of information and communication technologies (ICT) in working with families of early aged and preschool children with visual impairments. The authors analyze the current situation in families during strict quarantine in 2020 and 2021 and study the possibilities of using information technology in the development of children with visual impairments at home. The study describes how or under what circumstances specialists worked with families of early aged and preschool children with visual impairments, what problems parents, professionals faced, and whether they were ready to solve—via the author's questionnaires for professionals and parents, identified current problems that need urgent solutions. The theoretical basis and practical side of studying the problem of using information and communication technologies in families' distance psychological and pedagogical support are presented. Attention is focused on different levels, and specialists with children with visual impairments in the context of family education and the absence or insufficiency of distance correction work with early aged children and their parents are emphasized. The need for further study of the situation for the development of appropriate forms, methods of ICT in working with families of children of early and preschool age with visual impairments. The relevance of scientific research on the use of ICT and distance psychological and pedagogical support is proved, as this format will be in demand not only during lockdown but also in seasonal outbreaks, challenging weather conditions, as well as constantly for families living in settlements, remote from International Rescue Committee and other educational institutions, families who due to various life circumstances are not able to visit institutions that provide educational and psychological services. It is concluded that there is an urgent need to create a holistic system using ICT to work with families of early aged and preschool children with visual impairments, which will help solve a wide range of psychological and pedagogical rehabilitation support tasks. Keywords:

* Corresponding author: tanya_t.k@ukr.net

information and communication technologies, children with visual impairments, early age, preschool age, distance learning, surveys.

1 Introduction

The modern period of educational development is characterized by the increasing role of information and knowledge, which become factors of social progress and well-being. The development of the information society causes changes in almost all spheres of life. The availability of information and communication technologies (ICT) dramatically changes the way we work with data, simplifies the processes of knowledge acquisition, exchange of experience and interaction between people.

Ukrainian society is currently experiencing severe challenges related to the periodic imposition of quarantine in the country caused by the effects of the COVID-19 pandemic. In such crises, professionals working with children with visual impairments show an increase in problems of various kinds due to the low level of information and communication technology during the period of remote provision of counselling, correctional and diagnostic services. Psychological support for families of children with visual impairments is usually provided in the absence of ICT support and low correctional and diagnostic capabilities. These facts lead us to say that issues related to providing psychological assistance to families of visually impaired children who are in solitary confinement can no longer remain within the scope of providing individual assistance to parents or children. There is an urgent need to develop a remote psychological and pedagogical support system for families with children with special educational needs in modern Ukraine, subject to a prior detailed study of the issue.

Formulation of the problem. The lack of a systematic approach to creating opportunities for ICT use in working with families of young and preschool children with visual impairments has become a serious obstacle to effective care for these families. In conditions of self-isolation due to a pandemic, children with visual impairments do not receive proper exceptional and psychological and pedagogical support, which negatively affects their development, inhibiting the success of habilitation, rehabilitation and inclusion.

Currently, scientists from the Mykola Yarmachenko Institute of Special Pedagogy and Psychology of the National Academy of Pedagogical Sciences of Ukraine have developed distance learning models for people with special educational needs in quarantine, which allow them to work at home with the whole class under the guidance of a teacher—mastering the methods of learning. Many measures are taken to train specialists in providing psychological assistance to families raising children with special needs: special seminars and supervision groups are held for psychologists, social educators and representatives of other related professions; methodical materials of informative and educational character are developed and distributed (mainly through social networks), etc. However, an essential condition in working with visually impaired children in correctional and developmental work should begin as early as possible. However, today not only there is little research on the education, upbringing and socialization of children with visual impairments of early and preschool age, but also no work and experience on the use of ICT in remote support for children in this category and their parents [1], [2].

Therefore, the issue of development and implementation of scientific research, which will determine the organizational and methodological conditions for using ICT in working with families of visually impaired children of early and preschool age, is acute.

ICT in working with families of visually impaired children of early and preschool age in the remote form of psychological and pedagogical support is an innovative form of support based on both the best traditional methods of support and new information and interactive technologies. At the same time, this form is not available to all population segments and depends on many factors, including material security, place of residence and health status [3].

2 Analysis of recent research and publications

A number of domestic and foreign studies have focused on addressing the introduction of ICT in the educational process of children with special needs, taking into account their educational needs, in particular in the works of B. Eckert, G. Hoppe, H. Lobin, V. Bykov, V. Zasenko, B. Moroz, V. Ovsyanyk, L. Prokhorenko, O. Rybalko, O. Romanenko, B. Shunevych and others. [4], [10].

Researchers worked on the development and application of practical forms and methods of working with children with visual impairments: V. Bochelyuk, O. Bezpal'ko, I. Hryha, E. Daniels, N. Zhyhaylo, I. Zvyeryeva, I. Ivanova, A. Kaps'ka, A. Kolupayeva, H. Laktionova, M. Lukashevych, I. Myhovych, T. Semyhina, P. Talanchuk, S. Tolstoukhova, A. Turubarova, L. Tyuptya, M. Firsov, O. Kholostova, M. Chaykovs'kyy, Ye. Yars'ka-Smyrnova and other scientists.

Problems of life of children with visual impairments are considered in research H. Hryhor'yevoyi, M. Zemtsova, N. Kyyanytsi, Z. Kruchynina, K. Lebedyns'koyi, V. Myasyshcheva, L. Hovikovoyi, Ye. Syn'ovoyi, L. Solntsevoyi, V. Feoktystovoyi, I. Kholkovs'koyi, V. Shnayder, D. Uorrena. Along with these studies, there are several works on the effects of visual deprivation (V. Andriyenko, L. Vavina, L. Vyhot·s'kyy, V. Zhurov, V. Kobyl'chenko, L. Kunenko, O. Lytvak, I. Morhulis, L. Nafikova, V. Remazhevs'ka, T. Svyrydyuk, Ye. Syn'ova, Ye. Sternina, A. Suslavichus, V. Feoktystova etc.). The authors argue that in conditions of visual deprivation disrupts the processing and perception of the information content of cognitive material by children, so the organization of the educational process requires special additional equipment and informatization [6], [7].

Currently, the informatization of the education system relies on a set of theoretical and methodological provisions in the education of children with special needs: the use of information technology in special and inclusive education is found in the works V. Zasenko, B. Moroz, V. Ovsyanyk etc.; research of possibilities of increase of efficiency of training of such children with the use of electronic game resources in elementary school - V. Bikov, S. Litvinova, O. Melnik; study of the peculiarities of the classification of objects of the information field by children with mental retardation - L. Prokhorenko, O. Romanenko [2], [8], [11].

However, despite differing views on this issue, researchers agree that using information and communication technologies will contribute to the acquisition of skills to apply knowledge in a variety of, sometimes extraordinary, situations, enriching the practical experience of a child with special educational needs.

A child with visual impairments has the opportunity to learn about the world around them only with the help of preserved analyzers. Therefore, many researchers believe that the mental development of such children, the formation of compensatory processes, active life position, awareness and mastery of ways of self-realization depends on various conditions - biological, social, psychological, and incredibly educational (T. Dehtyarenko, K. Dovhopola, Ye. Klopota, V. Kobyl'chenko, O. Lehkyy, Ye. Syn'ova). Scientists claim that the use of information technology in the education of children with visual impairments creates favourable conditions for their education in all educational institutions, regardless of ownership, access to educational services, active participation in society and others.

Unresolved aspects of the problem. Despite numerous scientific studies, the use of ICT in working with families of young and preschool children with visual impairments remains open and underdeveloped. According to scientists, the success of the education of children with visual impairments depends on the educational environment in which the child is, on the correctional and developmental orientation of the content of education, which should include the development of compensation, correction and restoration of impaired functions, search for potential opportunities for development and knowledge children with visual impairments. In this context, the very use of ICT allows children to learn according to their characteristics and capabilities, even remotely.

The purpose of the article is to outline the results of studying the use of ICT in working with families of young and preschool children with visual impairments in the context of remote psychological and pedagogical support.

3 Theoretical fundamentals of research

The analysis of theoretical research shows that the issue of using ICT in working with families of young and preschool children with visual impairments became acute when educational institutions were forced to switch to distance learning. During this period, the organization of the educational process for children and the continuation of counselling, correctional and developmental work with children with visual impairments at home was threatened. Given that ensuring access to education for children with visual impairments requires the use of special devices, which are one way to compensate for visual deprivation, a channel of communication with the environment, a conductor of information, the question of studying ICT in working with families of children with visual impairments early and preschool age [8], [9].

In our opinion, the development and application of flexible strategies and techniques, using modern information and communication technologies that would take into account the individual characteristics of families with young and preschool children with visual impairments, will address the above issues of further remote psychological and pedagogical support.

It should be noted that ICTs act as a powerful means by which specialists in different situations can always accompany a child with visual impairments. The use of information technology promotes the development of children's visual impairment of thinking, memory, ability to self-organization and co-creation, increases learning motivation, and hence the level of their socialization [5], [9].

Scientists conclude that information technology has a positive effect on the process of learning and education primarily because they change the scheme of knowledge transfer and teaching methods [2], [5]. At the same time, the introduction of such technologies in the education system of children with visual impairments is based on the use of special equipment, software and hardware, information processing systems, unique teaching methods and more. There is very little research on ICT use for children of this nosology of early and preschool age, which requires complex and active intervention from scientists.

4 Research methodology

The experimental work was carried out in three stages. In the first stage, questionnaires with open-ended and closed-ended questions were developed for professionals working with visually impaired children and for parents, grandparents, guardians or other persons directly involved in the care, development and upbringing of visually impaired children from birth to the age of 7-8 years at home.

The purpose of the survey was to find out the current situation in the families and communities of professionals during the severe quarantine in 2020 and 2021, the possibilities of using information technology in the development of children with visual impairments, to identify current problems that need urgent solution.

The second stage involved the distribution of questionnaires on social networks, data collection. At the third stage, all the obtained data were processed and processed, including Fisher's angular transformation that had been used.

5 Research results

One hundred fifty-seven answers were received to the questionnaire for specialists, including 56.7% of teachers, 30.6% of correctional teachers, 9.6% of psychologists, 3.2% of specialists said they combine several functions, 3.2% - chose the answer " Your option. " It should be noted that none of the specialists chose the item "early intervention specialist", which, in our opinion, is a confirmation of the fact of acute shortage of such specialists in Ukraine.
Among the respondents, 17.2% work with young children and 82.8% with preschool children. At the same time, 11.5% of all specialists deal with children of both early and preschool age.

Of all respondents, 84.1% during the lockdown in 2020 worked with children with visual impairments in the online format using ICT, respectively, 15.9% did not. This indicates that most professionals quickly (not to say that comfortable and/or practical) moved to a new work format. In our opinion, this is because the education of children with special educational needs in Ukraine is currently under development and reform. Most professionals are constantly looking for new technologies and work methods, attend various training events and are open to new things, so they are potentially ready to react quickly to change.

However, it should be noted that only 51.6% of respondents conducted online classes (of various formats) with all their students (clients), while 48.4% of professionals had families with whom there were no relevant classes. The reasons for this difference can be different, for example, the unwillingness of families to work in a new format for them, lack of technical capabilities and others.

To the question "What format of work did you use?" we received the following answers (note that one respondent could choose several answer points if he used different forms): 74.5% conducted online consultations for parents (or persons caring for the care, upbringing and development of children with visual impairments); 50.3% conducted online classes directly with children; only 10.8% disassembled home video and provided relevant advice on its results; 73.2% sent tasks that parents performed independently at home with their children; 70.1% provided telephone consultations. Separately, we note that 4.5% of respondents did not use the remote format, and 12.1% chose the item "other". Note that the last two indicators are close to 15.9% of professionals who did not work with children with visual impairments online. At the same time, under the item "Other", there may be such options as "met on the street during a walk", "gave home visits", "took the child home for private lessons", or others.

Among the respondents, 89.8% provided assistance (of various formats, not only with the use of ICT) to families and children during the severe quarantine and lockdown in 2020. Only 10.2% did not work with their families at that time.

When asked whether the situation with the use of ICT has changed during the lockdown in early 2021, 56.7% answered in the affirmative and 43.3% - in the negative.

The following answers were received to the question of what was the biggest obstacle and challenge in the organization of work with the use of ICT (respondents could give several answers): 61.1% indicated the lack of material and technical base or difficult access to it (in including poor Internet connection) by specialists; 27.4% lack of methodological assistance on working with visually impaired children of early and preschool age in the online format; 76.4% lack of direct contact (including tactile) of the child with a specialist; 17.8%

psychological barriers they felt due to the need to move to a new format of work; 68.2% noted the unwillingness of families to do such work; 3.8% chose the item "Other". Thus, experts consider the biggest obstacles to the lack of direct contact with the child and the unwillingness of families to move to work with ICT. It should be noted that the training, seminars, and the availability of appropriate educational and methodological literature could help overcome these problems and the psychological barriers of the specialists themselves. As for the lack of sufficient material and technical base, it is impossible to solve it within the framework of psychology or pedagogy; this issue should be taken care of by the relevant structures, including and public.

It is interesting to compare the answers described above with the answers to the questions about what was the biggest obstacle and challenge in organizing work using the latest technologies for families of visually impaired children (according to experts): 61.1% noted the lack of material and technical base or complicated access to it (including poor internet connection) for families; 65.6% lack of free time for parents (or persons who replace them); 43.9% of psychological barriers experienced by parents during the transition to the new format; 9.6% chose the item "Other". Thus, we see that families also had many barriers to implementing the educational process through ICT, some of which can be overcome within the framework of psychology and pedagogy, and some - no.

According to the interviewed experts, families growing up with children with visual impairments need the most help during the lockdown in the following areas (meaning those issues where the support of a teacher or psychologist may be relevant): 78.2% of assistance in organizing developmental, correctional and training classes; 40.4% of consultations on the organization of everyday household issues for the care and upbringing of children; 30.8% assistance in maintaining a favourable microclimate in the family as a whole; 7.7% of other assistance.

To the question of what the personally interviewed specialist lacks the most in conducting consultations and classes using ICT or increasing their efficiency, the following answers were received: 68.8% need to improve the material, and technical base (including high-quality Internet communication); 68.8% lack the desire and willingness of families to work online; 35% would like to receive scientific and methodological support; 22.3% feel the need for psychological support from the educational community (supervision, exchange of experiences with colleagues, etc.); 12.7% overcoming one's psychological barriers and more self-confidence; 6.4% chose the item "Other".

When asked about the prospects for further development and use of ICT in providing consultations and classes with families of visually impaired children, 75.8% answered in the affirmative, 24.2% of respondents do not consider this format promising. In our opinion, this aspect of the study needs to be studied in more detail to determine why experts consider a significant obstacle to the use of ICT in working with families and children and the possibility of overcoming them.

It should also be noted that in response to an open-ended question about what worries professionals in the context of the use of ICT in working with visually impaired children, many of them noted that the issue of preserving children's vision is acute, as well as the lack of conferences (seminars, lectures) including online for parents and professionals who have had to adapt quickly to change.

In parallel with the questionnaire for specialists, a questionnaire was opened for parents (or persons who replace them). One hundred eighty-two people answered this questionnaire. Of these, 97.3% raise a child with visual impairments, and only 2.7% are blind. By age, the answers were divided as follows: 0.5% of the child's parents from birth to 1 year; 7.7% from one to three years; 24.7% have a child of the fourth year of life; 33.5% of the fifth year of life; 33.5% raise a 5-6-year-old child, 33.5% 7-8-year-old. Thus, we can see that the activity of parents and their need for training with a specialist increases as the child grows older.

At the same time, 51.6% of respondents answered in the affirmative whether a child receives specialized psychological and/or pedagogical assistance in a child care centre, IRC or otherwise, and 48.4% do not receive such assistance. Thus, we see that the coverage of specialized care for children of early and preschool age with visual impairments is currently insufficient. About half of the children in this category and their families do not receive help from a pedagogue or correctional teacher, a particular psychologist, etc.

Among the surveyed parents, 47.8% experienced increased anxiety, helplessness, tension during the lockdown. Moreover, 58.2% noted that the family had additional difficulties in raising and developing the child. Among the difficulties were noted, as problems of material and technical nature and organizational, psychological. Many parents complained about the lack of free time (as mentioned by experts), increased anxiety in children and parents, increased hysteria of children, poor quality of distance learning and the difficulties experienced by children during such activities, lack of live communication more. A special mention should be made of the deterioration of children's well-being associated with the vision strain, including redness and pain in the eyes. All these points indicate that the families of children with visual impairments are in dire need of informational and psychological support in several issues, including the organization of classes and leisure with the use of ICT.

When asked whether, during the hard lockdown in 2020, the family received specialized psychological and/or pedagogical support in a remote format, 52.7% of respondents answered in the affirmative. Accordingly, 47.3% did not receive such support.

Among those families who received specialized psychological and/or pedagogical support in a remote format, the most common (respondents were able to choose several items): performing exercises and tasks sent by a specialist (45.1%); online classes with children (34.1%); telephone consultations (25.8%); online consultations of parents (22%); analysis of home video and further consultations on its results (9.3%); 3.3% chose the item "Other".

Among those families who worked with ICT professionals, 43.5% considered this format effective for their child, 43.5% could not decide on the issue, and only 13% believed it was inadequate for their child. We see that a significant proportion of parents and children have specific difficulties and obstacles that hinder the full implementation of the academic press through ICT. The issue of overcoming them needs further study.

Among the respondents, 51.1% said that in 2021 the situation with receiving specialized psychological and/or pedagogical support in a remote format during a hard lockdown has changed for the better; respectively, 48.9% did not notice such changes. Note that these figures do not differ significantly from the answers of experts to the same question.

When asked whether parents would like to receive specialized psychological and/or pedagogical assistance in remote and/or face-to-face formats using ICT in the future, 56.6% answered in the affirmative, 31.1% have not yet decided, 12.1% have not would like this. Thus, most parents see the prospects for their children to use ICT in an educational format, but the share of those who have doubts and fears is also relatively high.

According to the respondents, most families with visually impaired children need help in the following areas during the lockdown (meaning the issues where the support of a teacher or psychologist may be relevant): 77.5% of assistance in organizing developmental classes; 22.5% of consultations on the organization of everyday life, care and upbringing of children; 18.1% assistance in maintaining a favourable microclimate in the family as a whole; 16.5% chose the item "other". We see that the vast majority of families with children with visual impairments are in dire need of help on many issues, so the urgency of organizing such care.

We compare these answers with the answers of specialists. Comparing the data of both groups on the need of families to help organize developmental classes (specialists 78.2%, parents 77.5%), we see that there is no significant difference between the two groups in this issue ($\varphi^* = 0.046$).

Instead, regarding the need of families for consultations in the organization of everyday life issues, care and upbringing of children (specialists 40.4%, parents 22.5%) and the need to help maintain a favourable microclimate in the family as a whole (specialists 30.8%, parents 18,1%), the opinions of specialists and parents on this issue differ ($\varphi^* = 3.516$ and $\varphi^* = 2.69$, respectively). The proportion of professionals who believe that families need help in both matters is greater than that of parents who think the same. This can be due to several reasons; for example, specialists do not know and understand the daily life of families with children with visual impairments; family members consider their daily life and family microclimate normal, acceptable, and the expert has a different opinion on this, believes that daily rituals and affairs, family communication needs to change for the better, and so on. It is also possible that there is no close and open contact between families and specialists. There are no honest discussions of failures and successes. This issue needs a separate study.

According to the interviewed parents, the biggest obstacle and challenge in organizing the use of ICT for the family were: 65.9% noted the lack of direct contact (including tactile) of the child with a specialist; 32.4% lack of material and technical base or difficult access to it (including poor Internet connection); 20.3% psychological barriers they experienced during the transition to the new format; 17% unwillingness of specialists to work in a remote format; 13.7% chose the item "other".

Comparing the answers of both groups regarding the lack of direct contact (including tactile) of the child with the specialist as the biggest obstacle (76.4% of specialists, 65.9% of parents), we see that the sensitivity of direct contact between the specialist and the child has become a bigger problem. Specialists ($\varphi^* = 2.139$), although a significant proportion of parents noted it. In our opinion, the significance of this obstacle for specialists is determined by the peculiarities of working with children with visual impairments and the age characteristics of early and preschool age. The difference could also be influenced by the fact that the specialists did not directly contact the child while the parents had it.

Difficulties caused by the lack of material and technical base or difficult access to it (including poor Internet connection) for families were noted by 61.1% of specialists and 32.4% of parents. We can see that among the first, a significantly higher proportion of those who believe that this has become a severe obstacle for families to study using ICT ($\varphi^* = 5.362$). This difference in shares can be explained by the fact that a significant number of families with difficult access to the Internet were unable to participate in this survey.

There are also differences in views on the psychological barriers experienced by parents during the transition to the new format (43.9% of professionals, 20.3% of parents). Experts were much more likely to note psychological barriers in families ($\varphi^* = 4.71$). Several reasons can explain this difference: for example, people often tend to evaluate themselves and others differently, even on the same issue. It is also possible that experts focused on working with children and could not accurately assess the condition and readiness of families to work with ICT. It is also permissible to lack complete contact and, accordingly, honest discussion of the problems in both respondents' groups during distance learning. The issue of complete contact requires further research to determine the causes.

Among the surveyed families, the following formats of remote work are considered the most effective for their child: 54.1% noted online lessons using ICT; 34.3% video recording with further analysis and consultation with a psychologist (teacher); 22.1% of telephone consultations; 19.3% fulfilment of tasks of a psychologist (teacher) without the use of feedback through ICT; 18.2% of receiving assignments and writing reports via e-mail; 13.3% chose the item "other".

Among the answers to the open-ended question about what worries parents the most about this topic, it was expected to be mentioned: fears of reduced visual acuity due to the use of gadgets, inability to obtain specialized medical procedures prescribed by an ophthalmologist; lack of special training sessions for specialists and parents on the organization of online training.

6 Conclusions and prospects of further research

Based on the analysis of recent research and publications, it can be concluded that the relevance of the use of ICT for remote psychological and pedagogical support of families with children with special educational needs is growing due to global changes in the world and in Ukraine. At the same time, such a category of children as children with visual and preschool visual impairments, their families and professionals who work with them is insufficiently studied in this context.

We can also conclude that children with visual impairments of preschool age are not fully covered by specialized psychological and pedagogical assistance. The lack of such care is especially acute for young children.

During the lockdowns in 2020 and 2021, not all families with preschool children received psychological and pedagogical assistance using ICT. It should also be noted that the transition to a remote format of work was somewhat easier for specialists than for parents. However, both groups had several problems and obstacles along the way. These problems were both material and technical, as well as organizational and psychological. Not all identified problems can be overcome within psychology and pedagogy. Some of them need to be decided at the state level to solve them.

Most experts and parents believe that using ICT for remote psychological and pedagogical support of families raising children with visual impairments is promising. Note that parents have more doubts about this issue than professionals.

The creation of a system of assistance to both parents and specialists in organising remote psychological and pedagogical support for children with visual impairments is acutely relevant. It is necessary to develop, test and implement general approaches and individual methods of educational and, in particular, a correctional process using ICT. Particular attention should be paid to reducing the difficulties in this category of children and professionals who work with them, which arise due to the lack of direct (including tactile) contact.

It should be noted that this study shows that further study requires a number of issues, in particular: the issue of establishing interaction between professionals and parents during the remote format of work; identifying and overcoming psychological barriers in the use of ICT in both groups; preservation of vision in children and prevention of its disorders, as well as prevention of general psychological exhaustion in children, parents and professionals during remote work and isolation; research to increase the effectiveness of the use of ICT in working with families raising children with special educational needs, etc.

It should be noted that scientific research on the use of ICT and remote psychological and pedagogical support, in our opinion, is relevant and promising because this format of work will be in demand not only during the lockdown but also in seasonal outbreaks, challenging weather conditions, as well as permanently for families living in settlements remote from the Irish and other educational institutions. Due to various lifestyles circumstances, these families do not allow to visit institutions that provide educational and psychological services.

References

1. O.O. Babiak, L.I. Prokhorenko, N.I. Batasheva, Visnyk natsionalnoi akademii pedahohichnykh nauk Ukrainy, **1(2)** (2020), https://doi.org/10.37472/2707-305X-2020-2-1-7-5 (in Ukrainian)
2. V.Yu. Bykov, S.H. Lytvynova, and O. M. Melnyk, Informatsiini tekhnolohii i zasoby navchannia, **6(62)**, 34-46 (2017), https://journal.iitta.gov.ua/index.php/itlt/article/view/1937/1289 (in Ukrainian)
3. Brandon Hall's research data. http://www.slideshare.net/msquareg/cms-lms-lcms
4. V.V. Zasenko, *Use of information technologies in the conditions of special and inclusive education of children with hearing and speech disorders*, http:rc-vaboc.at/ ua/_ld/0/9____________.pdf (in Ukrainian)
5. Ye.A. Klopota, Aktualni problemy navchannia ta vykhovannia liudei z osoblyvymy potrebamy, **12(14)**, 142-155 (2015) (in Ukrainian)
6. V.V. Kobylchenko, Pedahohika i psykholohiia, **2**, 64-70 (2016) (in Ukrainian)
7. T.M. Kostenko, V.I. Hoshovskyi, Osvita osib z osoblyvymy potrebamy: shliakhy rozbudovy, **14**, 91-95 (2019) (in Ukrainian)
8. L.I. Prokhorenko, O.V. Romanenko, Informatsiini tekhnolohii i zasoby navchannia, **74(6)**, 266-277 (2019) https://www.researchgate.net/publication/338334638_OSOBLIVOSTI_KLASIFIKACII_OB%27EKTIV_INFORMACIJNOGO_POLA_DITMI_IZ_ZATRIMKOU_PSIHICNOGO_ROZVITKU (in Ukrainian)
9. R. Simpson, *Behavior Modification and Child Management*, in Handbook on Parent Education, N.Y., USA, 153-158 (1980)
10. A. Talley–Organ, *Typical and atypical development in early childhood the fundamentals,* Allen Unions, USA (1998)
11. G. Tudg, *Collaboration, conflict and cognitive development: the efficacy of joint problem solving*, in Proceedings of the Eastern Psychological Conference, 18 April, N.Y., USA (1986)

The Gravity of Cryptocurrency and Prospects in a Post-Pandemic Economy

George Abuselidze[1*], and *Gia* Zoidze[2]

[1]Batumi Shota Rustaveli State University, 6010 Batumi, Georgia
[2]Batumi State Maritime Academy, 6010 Batumi, Georgia

Abstract. Today, everything in the world is changing rapidly; businesses that ten years ago may not have existed at all now exist and bring with them a significant income. Change in almost every field has one key feature: it is digitalisation. There are innovations in payment methods as well; the most high-profile issue in recent years in this regard is the emergence of cryptocurrencies. Within a few years of its appearance, it has become popular not even locally but worldwide. The method of data processing and analysis, quantitative and qualitative research methods, statistical-economic analysis methods are mainly used when working on the paper. The research aims to identify the main directions of cryptocurrency development in globalisation threats and challenges. Most cryptocurrencies are built on blockchain technology. The technology of the future is designed to securely store information, expedite transactions, reduce international transaction costs, replace cb-money with e-money and eliminate intermediary circles. Keywords: Money, Digital Currency, Cryptocurrency Market, Energy Deficit.

1 Introduction

What is cryptocurrency? - Nowadays, many people are looking for the answer to this question. However, before answering this, let us ask a more straightforward question: What is money? If you convince those around you, money can become anything, an object, or even an idea without a physical body. Money can be anything that can perform the function of money.

As a unit of trade and exchange, money has existed since ancient times, but many things have changed since the development of the economic process and the emergence of new ways of communication. The last point of these changes was the emergence of cryptocurrency, which, among other things, we must "blame" for technological progress.

The idea of creating e-money appeared years ago, but the first successful embodiment of this idea was Bitcoin. For many, this industry is associated with financial speculation, while others see the future of economic and financial relationships only in the digital world.

However, what is cryptocurrency? It is a digital currency that uses any transaction to record and store in a decentralised system known as a blockchain. Any information executed in a cryptocurrency in a blockchain is constantly updated, and it becomes

* Corresponding author: george.abuselidze@gmail.com

impossible to falsify it. In order for the transaction to get into the blockchain, miners are needed.

The working principle of the miners is simple: they load new information into the block, and in return generate cryptocurrency, which can be considered as a reward for the work done. However, no matter how we delve into the details, the major revolutionary step in shaping the crypto world is integrating this level of money into the digital world, thus creating a whole new, decentralised financial reality that has yet to unleash its full potential [1].

A clear example of the new financial reality is the financial exchanges where cryptocurrency trading takes place daily. While there are still no outlines of how this area can be regulated at the legislative level, the fact is that the number of people expressing a desire to buy and trade cryptocurrency is growing daily.

Technological progress dramatically simplifies this process, which is why we have to put up with the reality we have encountered so quickly. It is impossible to ignore this reality, and we must find ways to adapt because when it all comes to the surface and touches every detail of our daily lives, then it may be too late. Therefore, we must keep pace with technological progress and use it to our advantage - to create a new, future reality [2].

2 Literature Review

The paper is based on the theory of analysis and synthesis of concrete examples of international and Georgian experiences. Key issues are resolved on the basis of control and functioning relationships.

We can mainly highlight the works of the following authors: A. Bailey (2019), E. Bouri, S.J.H. Shahzad and D. Roubaud (2019), C. Brenig and G. Müller (2015), D. L. K. Chuen, L. Guo and Y. Wang (2017), A. E. Bahrawy, L. Alessandretti, A. Kandler, R. Pastor-Satorras and A. Baronchelli (2017), G. Hileman and M. Rauchs (2017), M. Iwamura, Y. Kitamura, T. Matsumoto and K. Saito (2019), Y. Liu and A. Tsyvinski (2021), U. A. K. Rogoff (2021), E. Tukhiashvili (2021), et al. [2-13]. However, despite the multifaceted discussion of the problem of cryptocurrency management (especially in conditions of uncertainty), it has been improperly studied and researched in the international literature and scientific publications. Bailey's paper analyses that one of the reasons why Bitcoin is among the top ten cryptocurrencies is that it is the oldest it has the largest community of developers and investors, which contributes to its further growth [2]. Authors showed that Most of the restricted prove on the exponential cost spikes (i.e. cost explosivity) within the cryptocurrency showcase considers the case of Bitcoin, even though other cryptocurrencies have slowly dissolved Bitcoin's dominance [3]. Against the background of global processes, authors determine the mechanism of functioning of the international currency market in the wake of its historical development [6, 14-17]. The authors discuss the essence and features of cryptocurrency, modern types of electronic money and blockchain technology, central bank regulation mechanisms, and the e-wallet and its characteristics, dynamics of cryptocurrency types (especially bitcoin regulation) and cryptocurrencies [8, 18].

According to the Authors, If not for the e-currency market, which would link economic entities in Georgia with foreign markets, entities in Georgia would have little chance of engaging in international currency trading. Through the electronic currency market, corporations and citizens of Georgia can acquire the necessary knowledge and experience, to be constantly aware of the current processes in the world currency market. At the end of the paper, based on the grouping of facts, combining them with common characteristics and their analysis, the main conclusions and recommendations are presented [19, 20].

3 Results and Discussion

3.1 Money in the Past, Present and Future

Everything may be utilised as money if people agree and the right purpose is assigned to it. Money must fulfil multiple requirements: it must be a payment method, a means of collecting/storing value, a means of accounting, and it must be easy to use.

Money has existed for a long time as the equivalent of universal interchange, and the notion of economic interactions has led to its progressive evolution. Since the dawn of time, people have agreed that to carry out the trade process with one another, something universal, a standard that could be swapped for anything, was required [4].

The global equivalent unit was formed and evolved in this way. With the increasing degree of growth in society, this unit has been continually modifying at different periods of human evolution. Wheat, products, gold coins, currencies, and even cryptocurrency were all used at various eras. Cryptocurrency is the most recent technical advancement in the field of money.

There are over 5,000 cryptocurrencies in circulation throughout the world. Among them, Bitcoin ranks first (with a capitalisation of more than $ 1 trillion). Also in the top ten are Cryptocurrencies by Market Capitalisation: Etherium, Binance Coin, Cardano, Tether, XRP, Solana, Polkadot, USD Coin and Dogecoin (Table 1).

Table 1. Top Cryptocurrencies by Market Capitalisation and Current Price, (29.10.2021) [21]

№	Coin Name	Market Capitalisation	Current Price
1	Bitcoin	1.15 trillion	60 829.88
2	Etherium	447.41 billion	3 792.73
3	Binance Coin	78.22 billion	465.19
4	Cardano	70.7 billion	2.15
5	Tether	69.03 billion	0.99993
6	XRP	52.47 billion	1.12
7	Solana	47.95 billion	159.62
8	Polkadot	40.92 billion	41.43
9	USD Coin	32.75 billion	0.99987
10	Dogecoin	31.08 billion	0.24

Cryptocurrencies have a long history dating back to the creation of Bitcoin in 2008. Many attempts to establish worldwide electronic money existed before Bitcoin, but none were successful. Cryptocurrency appears to be conveyed by a mere arrangement of numbers and has no materiality. However, the situation is not so straightforward. Behind these transactions, in the case of Bitcoin, is a database - blockchain.

A blockchain may be compared to a decentralised database in which information is kept on several computers rather than a single control centre. Instead of a single control centre, data is disseminated among a number of computers that act as nodes. The copies on all of these computers (nodes) are identical. The nodes work precisely as the Bitcoin algorithm specifies. Trying to fabricate information here is impossible and meaningless since, despite the theoretical possibility of fabricating information on one node, information on other nodes throughout the network remains intact [6].

So the nodes are continually broadcasting to one other, communicating to each other, spreading information, and warning each other of changes in the blockchain. This data is synced between them regularly.

Let us take a look at what a blockchain. Each new block is appended to the preceding block using cryptographic procedures, making it impossible to modify the information stored in the prior block. This is how a lengthy chain of linked blocks is created, which is why this database is known as a blockchain. Miners, who are in charge of the system's functioning, records information or store it in transaction blocks.

The job of miners in the Bitcoin network is to validate all transactions. They get compensated in the form of newly minted bitcoins from the system in exchange for their efforts.

The whole bitcoin design is based on an incentive structure that encourages all players to be honest, rather than rewarding players (system participants) for dishonest behaviour, for which they would be condemned and held responsible.

Cryptocurrencies' whole design is built on the most challenging encryption and higher mathematics, with game theory utilised to develop successful consensus models. Throughout the bitcoin system, the structure is impartial and balanced, the game's rules are known to everyone and defined in the protocol from the start, nobody really trusts anyone at all, and good conduct is valued while negative behaviour is condemned.

Bitcoin is a game-changing invention, a technology that radically transforms the way people think about money. In our opinion, Satoshi Nakamoto (creator of Bitcoin) is a modern Prometheus who set fire to humans in his time. Satoshi lay the groundwork for a new era of decentralisation by demonstrating the paradigm of unregulated currency.

Organisations, enterprises, and people can no longer ignore the need for digital attitudes and technology transformation. Businesses all across the globe are preparing their existing processes, competencies, models, and transactions for digital transformation. This transformation has resulted in the digitisation of financial institutions and operations. Bitcoin, likewise, is a product of this technological advancement.

The public's interest in the perceived worth of this cryptocurrency has surged as a result of its rapid rise in value. As a result, many speculators and gamblers have hurried to invest in this currency in the hopes of achieving enormous future profits. This has not only acted as a key spur for the currency's perceived worth to grow, but it has also made the currency exceedingly unpredictable. Some traders consider bitcoins to be an appealing investment instrument because of the unique attributes of keeping the digital trace of individuals who deal in bitcoin anonymous and that it is not controlled by any governing body or firm.

On January 1, 2016, the sustainable growth of Bitcoin began. It started 2016 with $ 433 and ended with $ 959 - a 121% increase in value in 12 months. From January to December 2017, extra attention, the introduction of new cryptocurrencies, and media coverage increased the value of the cryptocurrency from January ($ 921) to December ($ 13,062). This record margin was ultimately unstable, and Bitcoin fell victim to its routine variability, which steadily eroded its previous gains. However, this only lasted until 2020. When economic activity slowed due to global isolation during the initial outbreak of the pandemic, the price of Bitcoin began to rise again (Figure 1).

Cryptocurrency started 2020 with $ 7,200. The pandemic and the policies pursued by the governments of various states have sparked investor fears about the global economy and accelerated the rise in the price of Bitcoin. At the end of November, Bitcoin was trading for $ 18,115. In December 2020, the price of Bitcoin reached $ 29,000, which means that from the beginning of 2020 to the end of the year, its price increased by 302%.

Bitcoin (BTC) in 2021 (from January to October 2021) has already grown by 100%. As early as February 21, 2021, Bitcoin (BTC) reached its highest point - $ 58,332. All this means that the most expensive e-money has grown by 100% in just two months since the beginning of 2021. At the beginning of January, the price of one Bitcoin (BTC) was $ 33,063, and as of October 2021, the price of one Bitcoin (BTC) is already $ 59,056 [22].

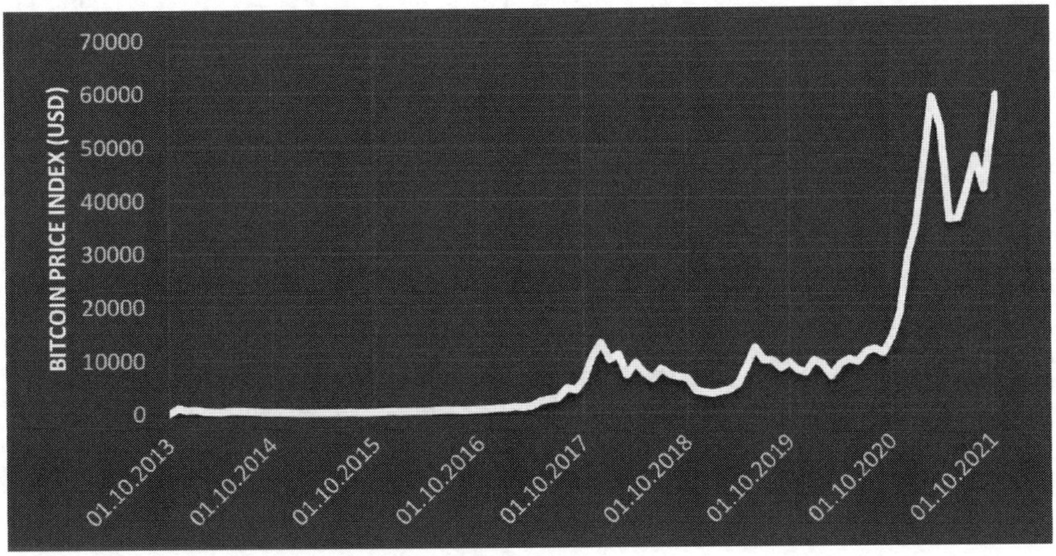

Fig. 1. Bitcoin Price Index in Dynamics, 2013-2021*

Source: Compiled by the authors based on data from [22]

At the same time, Elon Musk's company "Tesla" had a significant impact on the price of Bitcoin, after, on February 8, they invested a large part of the company's budget - $ 1.5 billion, in Bitcoin (BTC). It was announced that it would be possible to buy Tesla cars in cryptocurrency.

In this way, Tesla diversified its surplus assets in order to maximise profits. A few days before the information was released, Musk even posted #Bitcoin in his Twitter account description, indicating that he took cryptocurrency seriously. Analysts say Tesla's move will bring significant changes and shifts to the crypto world. At this point, the price of the cryptocurrency has risen dramatically.

It is also noteworthy that such a move from a company with so much influence was a bit unexpected, and it might give us a chain reaction. According to many experts, such a record rise in the price of Bitcoin (BTC) is due to the support for Bitcoin by the big companies that cryptocurrency has been receiving lately.

To buy Bitcoin, we first need to download the appropriate application (free of charge) and create a user that will allow you to get your account number. We must never forget the account number to complete the transaction, as our personal wallet is created with the account number. Also, bitcoin can be mined even if we do not complete any transactions - we just help to create blockchains and get cryptocurrency in it.

Nowadays, in many countries, it is possible to exchange the received bitcoins into the currency of your choice through both banks and ATMs. It is also possible to use it as a medium of exchange without exchanging for other currencies, as all of these cryptocurrencies currently have exchange rates.

3.2 The Main Challenges of Cryptocurrency Development in Georgia

Georgia, a country in Eastern Europe, is known in the crypto-society for two reasons:

- The first is cryptocurrency mining (due to its relatively low electricity costs, as well as its favourable tax environment and solid technical scene, which provide significant incentives for miners).
- The second is the fact that the National Agency of Public Registry has been recording land ownership information on the Bitcoin blockchain since 2015, which is a significant

tangible innovation in the public sector worldwide. This has been said and written about many times in the international press, which is good in itself and helps to position Georgia in a positive context.

In September 2018, the first international mining summit was held in the capital of Georgia - Tbilisi, where up to 1000 participants from different countries of the world gathered. Tbilisi has once again become the focus of the global crypto community.

Now mining is no longer so relevant because millions of people worldwide have decided to get rich in this simple way, buy a device and secure their network connection. Consequently, the production of cryptocurrencies has dropped significantly, and this activity has become less attractive. As for maintaining the image of innovation in the country, the initiatives of the Government of Georgia in connection with the 3rd generation blockchain - Cardano, are important in this direction. They announced cooperation with the Ministry of Education and Science of Georgia [11].

In another direction, the situation in Georgia is not so favourable. We do not have relevant regulations that will regulate the field. Georgia does not yet have the legislative clarity necessary for attracting foreign investors and developing the field. It is important to introduce appropriate regulatory norms in a timely manner. Estonia, for example, has managed to attract several billion in foreign investment simply because it has offered an appropriate friendly legal environment to investors (Estonia has so far issued up to 1,200 licenses for crypto e-wallets and trading platforms).

Most recently, an order of the Minister of Finance was issued in Georgia, which defines the issue of taxation of mining. However, there is still much to be clarified by the legislation. Unfortunately, an essential problem in Georgia is that Georgian banks do not show mercy to crypto companies; they do not serve them. Without banks, it is still impossible to produce full-fledged business processes and develop business [23].

The global pandemic has changed the structure of the economy. Consumer prices are rising catastrophically, the devaluation of the national currency continues, the countries are heavily dependent on imports, and there is no fundamental economic basis to stop these processes [24-27]. Over the last 2-3 years, the National Banks of many countries and international banks have been actively working on digital currency issues: conducting surveys, testing technical capabilities, and issuing digital currencies. One of the motivations for the release of digital currency is that if it is released properly, it will bring price stability to many sectors of the economy. Therefore, a stable digital currency for Georgia may be the most effective way out of this difficult situation.

It is the last year that cryptocurrency has decentralised financial resources concentrated in the hands of states. For example, if today the state's financial power is concentrated in its National Bank and it sets the rules of the game, the rapid development of cryptocurrency has brought about the fact that the state is losing influence in this area.
Consequently, in order to somehow stop this process and regain the reins of financial management, the central banks of many countries began to issue digital currency. By the way, small countries started it even earlier, five or six years ago, because they saw great investment potential in national cryptocurrencies [1].

Countries that did not have competitive advantages and additional economic strength, creating their own digital currency added investment attractiveness. An example of this is Malta, which has received more than $ 2 billion in foreign direct investment in the six months since the national digital currency legislation adoption.

Therefore, if we look at the current processes in the world, the entire global tax system is moving to cryptocurrencies. Georgia cannot be an exception.

3.3 SWOT-Analysis of the Cryptocurrency Market: A Theoretical Review

To recognise cryptocurrency as a means of payment worldwide, it is necessary to regulate its extraction, revenue and circulation processes, both in individual countries and around the world. At present, there is no common law in the world regarding the processes related to the regulation of cryptocurrency.

Proponents of digital currencies like to talk about the interest in virtual money of celebrities, sports and show business representatives; however, they stubbornly do not notice how much cryptocurrency has made life easier for hackers and extortionists. At the same time, anonymous cryptocurrencies have become an ideal tool for paying taxes, money laundering, organising terrorist attacks and other dangerous crimes.

The idea that cryptocurrency is a harmless way to accumulate is incredibly mimetic. Retailers typically make almost no use of virtual money due to its high transaction costs, but for anyone looking to avoid money laundering, money laundering, evading international sanctions, or committing many other crimes, they cannot find a better ally than cryptocurrency.

In order to better present the prospects and threats of cryptocurrency, we will try to present a SWOT analysis of the cryptocurrency market (Figure 2).

Sooner or later, governments will have to tighten their control over virtual money. Today, this process has begun, and as a rule, this should reduce the demand for cryptocurrency, but it still does not happen. So why is cryptocurrency expensive? The answer is given in part by economic theory, according to which giant soap bubbles appear on the market under conditions of zero interest rates. It is an asset that has virtually no value; however, so many large institutional investors have entered this market that politicians will find it very difficult to regulate cryptocurrencies [10].

In our opinion, the longer taking control of digital money takes time, the more difficult it will be to regulate cryptocurrency in the future. The fact is that much more international coordination is needed to control cryptocurrencies, at least in developed countries.

Bitcoin has become the driving force of the illegal global economy, and it will take decades for even the strongest and wealthiest countries to control it. The main and almost the only apparent advantage of virtual money is anonymity and uncontrollability. As soon as it falls within the framework of official regulation, the demand for it will decrease dramatically.

In addition, the development of the crypto industry requires huge amounts of electricity, which in all probability may lead to disruptions in the energy supply. It is a well-known fact that large cryptocurrency "farms" consume as much electricity as some cities. As a result, factories may slowly shut down, people may become unemployed, budgets may not be met, and costs may increase. Mass bankruptcy will be a matter of time.

In the event of the development of this scenario, humanity will have to deal with an invisible and, at the same time, a big problem - energy shortage. As a result, the entire burden of rising energy prices on the international market will affect the economy. Factories will stop, costs for farmers will increase, and almost all goods that require energy production will become significantly more expensive.

However, the crypto industry is also making progress in energy - making virtual money requires so much electricity that it will be possible to introduce new technologies in alternative energy in the world.

STRENGTHS	WEAKNESSES
▪ Time and Cost Savings ▪ Distributed Resilience and Control ▪ Open Source ▪ Auditability ▪ Security Features and Modern Cryptography ▪ Asset Provenance ▪ Native Asset Creation ▪ Dynamic and Fluid Value Exchange ▪ Higher Anonimity ▪ No Single Point of Trust ▪ Fraud Resistance ▪ Financial Incentives ▪ Division and Combination ▪ Predictable Rate of Coin ▪ Generation ▪ Non-physicality ▪ Decentralised Network ▪ Protected from Any Inflation ▪ State interference is excluded	▪ Currency in Name Only ▪ Security vs Usability ▪ Lack of Ledger Interoperability ▪ Lack of Hardened/Tested Technology ▪ Limitation of Smart Contract Code Programming Model ▪ Wallet and Key Management ▪ Lack of Trust in New Technology Suppliers ▪ Lack of Skilled Human Resources ▪ Immature Scalability ▪ Skills Scarcity and Cost ▪ Customer Unfamiliarity and Poor User Experience ▪ Braking anonymity through transaction history ▪ The deflation and hording problem ▪ Lost Bitcoins ▪ The transaction is not insured (usually you will not be able to get back the currency transferred after the transaction and no one will ever be able to return it to you) ▪ There is always a security risk (since e-wallets are one of the targets of hacker attacks. In addition, there is no state or international tool to protect cryptocurrency deposits)
OPPORTUNITIES	**THREATS**
▪ Active Software Development Community ▪ Business Process Acceleration and Efficiency ▪ Dependable Savings ▪ Reduced Transaction Costs (Decrease in Transfer Fees) ▪ Reduced Fraud ▪ Reduced Systemic Risk ▪ Monetary Democratisation ▪ New Business-model Enablement ▪ Application Rationalisation and Redundancy ▪ Gig Economy ▪ Libra as Onramp to Widespread Adoption ▪ Formation of Alternative Energy Sources	▪ Legal Jurisdictional Barriers ▪ Energy Consumption (Energy Management and Use) ▪ Systems Are Still Prototypes ▪ No Guarantee the Integrity of the Information ▪ High Complexity ▪ Politics and Hostile Nation-state Actors ▪ Speculative Bubble ▪ Data Privacy ▪ Money Laundering and Funding Terrorism Risks ▪ Institutional Adoption Barriers ▪ Technology Failures ▪ Divergent Blockchains ▪ Identification Through IP Address ▪ Identification Through The point of Sale/Exchange ▪ Heists and Robberies ▪ Poor Governance ▪ Ledger Conflicts/Competition ▪ Hacking Online Exchanges ▪ Possible Usage in Criminal Scenarios ▪ Volatility of Bitcoin Value

Fig. 2. SWOT Analysis of Cryptocurrency
Source: Compiled by the authors

4 Conclusion

In recent years, the facts of hacking trading platforms show that consumers should keep their cryptocurrencies not on trading exchanges but in a special crypto-wallet, the private key of which they will control themselves.

However, Bitcoin can be useful for many people. We can use it in any country, as it is an international currency. Blockchain is secure, and it allows us to make sure our money is moving/coming from the right person. Bitcoin recipients will not have to pay for transactions; Bitcoin has excellent support.

Sure, it has some downsides, but some of them happen because Bitcoin is a new, innovative product with high risks. However, as time goes on, the more the risks of uncertainty decrease and the less they pose a problem. All this will help Bitcoin gain more users, and if everyone uses Bitcoin, it can change the official currencies.

Given all the above, it should be emphasised that no one opposes blockchain technology, which is the basis of cryptocurrencies, and it has a huge potential for improving people's lives.

References

1. G. Zoidze, Three Seas Economic Journal **2(3)**, 1-7 (2021). doi:10.30525/2661-5150/2021-3-1
2. A. Bailey, The Future of Cryptocurrencies. KuppingerCole Report, ADVISORY NOTE (2019)
3. E. Bouri, S. J. H. Shahzad, D. Roubaud, Finance Research Letters **29**, 178-183 (2019)
4. C. Brenig, G. Müller, Economic analysis of cryptocurrency backed money laundering. (2015)
5. D. L. K. Chuen, L. Guo, Y. Wang, The Journal of Alternative Investments **20(3)**, 16-40 (2017)
6. A. ElBahrawy, L. Alessandretti, A. Kandler, R. Pastor-Satorras, A. Baronchelli, Royal Society open science **4(11)**, 170623 (2017)
7. G. Hileman, M. Rauchs, Cambridge Centre for Alternative Finance **33**, 33-113 (2017)
8. M. Iwamura, Y. Kitamura, T. Matsumoto, K. Saito, Hitotsubashi Journal of Economics, 41-60 (2019)
9. Y. Liu, A. Tsyvinski, The Review of Financial Studies **34(6)**, 2689-2727 (2021)
10. K. Rogoff, A Curse Worse than Cash. The World's Opinion Page: Project Syndicate. (2021)
11. E. Tukhiashvili, *Digital Georgian Lari - What will the cryptocurrency of the National Bank bring us?* Kviris Palitra. (2021)
12. G. M. Caporale, L. Gil-Alana, A. Plastun, Research in International Business and Finance **46**, 141-148 (2018)
13. Q. Ji, E. Bouri, C.K.M. Lau, D. Roubaud, International Review of Financial Analysis **63**, 257-272 (2019)
14. G. Abuselidze, Academy of Strategic Management Journal **18(4),** 1-10 (2019)
15. G. Abuselidze, N. Reznik, A. Slobodianyk, V. Prokhorova, *Global Financial Derivatives Market Development and Trading on the Example of Ukraine*, in Proceedings of the SHS Web of Conferences **74**, 05001 (2020). doi:10.1051/shsconf/20207405001

16. G. Abuselidze, European Journal of Sustainable Development **8(1)**, 53-68 (2019). doi:10.14207/ejsd.2019.v8n1p53
17. G. Abuselidze, L. Beridze, *The Role of Alternative Investments in the Development of Capital Markets in Terms of the Transformation of Georgia with the EU,* in Proceedings of the 4th International Conference on European Integration, 29-40 (2018)
18. Z. Ji, G. Abuselidze, V. Lymar, *Currency Internationalisation and Reforms in the Architecture of the Monetary System: the case of Digital Yuan*. Studies of Applied Economics. (to be published)
19. D. Katamadze, G. Abuselidze, G. Katamadze, A. Slobodianyk, Economics and Management **18(1)**, 19-32 (2021)
20. G. Zoidze, G. Abuselidze, Entrepreneurship **9(2)**, 33-46 (2021). doi:10.37708/ep.swu.v9i2.3
21. CryptoSlate, Coin Rankings. (2021). Available at: https://cryptoslate.com/coins/
22. Statista, Bitcoin price from October 2013 to October 29, 2021 (in U.S. dollars). (2021). Available at: https://www.statista.com/statistics/326707/bitcoin-price-index/
23. Business Media Georgia, Cryptocurrency - a new stage of finance. (2019)
24. G. Abuselidze, L. Mamaladze, Journal of Physics: Conference Series **1840**, 012040 (2021). doi:10.1088/1742-6596/1840/1/012040
25. G. Abuselidze, A. Slobodianyk, E3S Web of Conferences, **210**, 15016 (2020). doi:10.1051/e3sconf/202021015016
26. G. Abuselidze, *The Influence of Covid-19 on the Public Debt Growth and Default Risk: A Fiscal Sustainability Analysis,* in Proceedings of the International Scientific and Practical Conference on Sustainable Development of Regional Infrastructure - ISSDRI 2021, 151-159 (SciTePress, 2021). doi:10.5220/0010587501510159
27. G. Abuselidze, M. Sharabidze, MATEC Web of Conferences **342**, 08003 (2021). doi:10.1051/matecconf/202134208003

Priorities for the Development of Regional Information Management Based on System Digitalization

Yuliia Kovalenko[1], *Oksana* Zakharova[2*], *Oksana* Yakusheva[2,3], *Oleksandr* Yakushev[2,4], *Daniil* Hulak[2,5] and *Svitlana* Kozlovska[6]

[1]Donetsk National Technical University, 2, Shibankova Square, Pokrovsk, Ukraine
[2]Cherkasy State Technological University, 460, Shevchenko Blvd, Cherkasy, Ukraine
[3]Individual-entrepreneur Yakusheva Oksana Viktorivna, 46/20, Microdistrict Lazurniy, Pokrovsk, Ukraine
[4]Individual-entrepreneur Yakushev Oleksandr Volodumurovich, 13, Tolstoy str, Pokrovsk, Ukraine
[5]PJSC «Cherkasyoblenergo», Hoholya street, 285, Cherkasy, Ukraine
[6]Rauf Ablyazov East European University, 16, Nechuy-Levitsky str., Cherkasy, Ukraine

Abstract. The article is devoted to the substantiation of regional information management development priorities based on the digitalisation system. The paper substantiates the sequence of managerial actions in the development of Ukraine's economic digitalisation, taking into account existing changes in regional management, outlining the accumulated potential of science and practice, forming compliance with global progress of digitalisation and requiring a scientifically sound platform for regional information management. The directions of digitalisation intensification and information management development in Ukraine are formulated. The expediency of creating a specialised automated information system, the purpose of which should be to assess regional economic processes and prospects for their development in terms of economic growth, is substantiated. Requirements to information support of administrative activity in the system of regional management are formulated. The main tasks of information management support for the achievement of regional information management efficiency are defined. Keywords: Regional information management, Digitalization, Information and communication technologies, Economy of Ukraine.

1 Introduction

Acceleration of scientific and technological progress, strict adherence to innovative development is promising area of science. It forms the foundation of a scientifically sound solution to the problems hindering the movement towards European integration, the information economy and the information society in Ukraine. Solving problems and systemic conflicts of managerial, economic, socio-cultural nature is a prerequisite and factor in

* Corresponding author: bonheur5576@gmail.com

attracting breakthrough technologies and achieving the desired changes, compliance with the vectors of information progress, primarily to increase productivity; socialisation of the economy and its systemic cultural progress; introduction of innovative and information-intensive technologies.

With the use of information and communication technologies in all spheres of modern life, as well as with increasing quantity and quality of information, processing and assimilation of which is impossible without proper implementation of information technology, the question of forming a new, progressive vision of regional policy and regional governance. The formation of the information society requires a revision of state policy on regional development, development and implementation of state programs aimed at the priorities of information progress and digitisation of knowledge.

Society expects changes in the formats of the latest reforms, the continuation of the "old" reforms, the definition and strategic justification of growth points and vectors of priority movement. Modelling of the future is still active; processes have only just begun, but growth points have already been identified and the vector of the social economy and information society is emphasised. The scenarios of Ukraine's development focus on the systemic development of the IT sector, its state and multifaceted support; tax reform; mass privatisation; comprehensive de-shadowing of the economy.

2 The sequence of management actions in the development of digitalisation of Ukraine's economy

We will supplement and expand the vision of priorities for the development of Ukraine's economy based on systemic digitalisation through targeted processes and activities at the regional level and understanding the involvement of a full range of regional information management, vector for knowledge management in models of state, regional and sectoral management.

The phasing and scientific validity of such a movement is the primary interest of government and business, the need for time, an element of managerial progress. We recommend a sequence of management actions that consider existing changes, outline the accumulated potential of science and practice, form in line with global progress of digitalisation and require a science-based platform for knowledge management and regional information management.

The first stage. They define the role and place of regional information management in economic reform and compliance of the vector content of scientific provisions with the models of the information society; social, cluster and knowledge economy. The transition to the standards of the information economy, the mass phenomena of digitalisation are already actively taking place, the social economy and the knowledge economy appear to be developed models within the European space and need their optimal adaptation. The most significant and influencing the stimulation of the economy and society should be the transition to a cluster model (information-cluster; network-cluster) [1].

The second stage. Research and comprehensive analysis of changes in the economy of the Ukrainian regions in the direction of digitalisation. Independent analysis and identification of changes within the informatisation of government agencies and enterprises, industrial and market infrastructure; production modernisation; involvement and real efficiency of various information technologies; scale and comprehensiveness (territorial and sectoral) of modern information technologies.

Let's single out the changes that need their targeted research and will be recognised as key in further elaboration:

- determination of priorities, target vectors that correspond to the purpose of the trend of knowledge digitisation;

- recognition of fundamental changes characteristic of the latest information systems (information technology) support emergence;
- assessment of the available information potential and trends of its changes;
- evaluation of supply and demand in the market of IT services and IT products;
- assessment of technological support (technology, innovation, technology maintenance);
- development of knowledge management tools and technologies.

The third stage. Interconnectedness and assessment of interdependent processes of complex regional informatisation and corresponding changes in the economy, social sphere. Based on the achieved level of informatisation and automation, there is a rapid introduction of intelligent technologies.

Examples of smart cities that are actively working become practical embodiments of the most modern intelligent technologies. Such examples exist in Ukraine to a small extent. The city of Kyiv positions already tested and implemented smart technologies - e-ticket; video surveillance of car traffic in real-time; medical information services, etc., already downloaded to the Kyiv Smart City application. Access to such information is centrally published on the nashkiev.ua portal.

The fourth stage. Independent research and comprehensive study of the Ukrainian cities digitalisation processes from the standpoint of the future role of growth points in all regions (diagnosis, monitoring). Definition of zones and spheres of competitive advantages - information market, information services, the general process of regional digitalisation, the all-Ukrainian format of smartization of vital activity.

It should be noted that in Ukraine, the process of regions digitalisation is, relatively speaking, a "point". This means that computer and communication technology, telecommunications networks, databases and data banks, information technology, information systems and analytical centres of various levels, information technology products, systems of research institutions and training of highly qualified specialists first emerge and are implemented in large cities, and then diffusely spread into the region by district centres, etc. In addition, this process is more intense in regions with a high level of socio-economic development, thus causing the divergence of regions in the level of informatisation and further digitalisation in the form of intelligent technologies cooperation. Given that any process has limits of growth, sooner or later there will be a convergence of regions in terms of their informatisation, and hence in terms of information and technology support of the region's economy [2].

The processes of convergence and divergence manifest the informatisation intensification, the rapid advancement of smart technologies in the regions and the country as a whole, which covers all spheres of human life. Modern computer and telecommunication means are being introduced, computer networks are being laid, and databases are being created, the latest smart technologies of doing business are being formed, and the economy of information activity is developing significantly. At the same time, not all regions have the same opportunities to meet information needs based on the formation and use of information resources and modern technologies [3].

Fifth stage. Evaluation and study of regional issues of Ukrainian regions Internet coverage (target projects). Such research is systematically conducted by companies providing Internet services (PJSC "Ukrtelecom", PJSC "Datagroup", holding "Triolan", PJSC "Kyivstar", CC "Volya", PJSC "Vega", CC "Frigate", TeNet etc.).

Peculiarities of shifts in the regions of Ukraine covered by the Internet show an active growth in the number of users, the latest service technologies are being rapidly introduced. Regarding the territorial distribution, the most significant growth of Internet users is observed in the Central part of Ukraine - on average by 6.7% per year. In the Western region of the country, this figure is 4%. Unfortunately, the problematic military-political situation in the country has led to an annual Internet users number reduction by 6.5% in the East and 7.2% in the South [4].

Topical issues of Internet services market development need independent research:

- commercialisation of Internet resources in the regions of Ukraine. This question is relevant from the standpoint of expanding the boundaries of regional information markets and determining the business potential operating in the information market (and not only Ukrainian);
- promotion of smartization technologies and gaining experience in the relevant implementation of services (transport, education, medicine, etc.);
- cybercrime assessment;
- technical and technological problems of network access by regions.

The sixth stage. Assessment of asymmetry in the development of digitalisation processes at the level of the regions of Ukraine. The most relevant and popular are the issues of determining asymmetry:

- between cities and rural regions to attract information resources to make management decisions;
- gap in the minimum and maximum indicators of quantitative and qualitative changes (services, online services);
- in assessing the quality of IT services by region (for example, comparing the capital with peripheral cities);
- between the most successful and least equipped technologies and tools for digitalisation of regions.

The seventh stage. Obtaining a generalised diagnosis of the regions of Ukraine according to the main parameters of digitalisation (based on the target comparison with the indicators of Kyiv; European, for example, Poland).

The eighth stage. Formation of the targeted policy of digitalisation development in the regions of Ukraine on the platform of involvement and implementation of mechanisms and technologies of regional information management. Formation of a knowledge management model of regional information management taking into account economic, technical and technological, social and cultural interests of the population and business, conformity to the general movement of information progress.

3 Directions for intensifying the development of digitalisation and information management

Comprehensive movement in the direction of these stages involves the involvement of a large amount of digitised information and the latest knowledge. First of all, this is official information published on the official web portal of the State Statistics Service of Ukraine (www.ukrstat.gov.ua) and the main statistical offices of the regions. IDI indices developed and calculated by international organisations are relevant in the research (United Nations e-Government Readiness Index), IDI development index, network-ready index (NRI) and others. Only a comparison of data and results with world standards will provide an accurate picture of the promotion of integrated informatisation, digitalisation, e-government standards.

The driver, the main component of knowledge of regional information management and the corresponding transition to the targeted digitalisation model are information systems that ensure the transformation of information into solutions by the resource potential of government and various organisations, knowledge of people who operate and distribute information.

We recommend the initial vision of the classification of information technologies (applied), which are of strategic importance and form the potential for future deployment of regional information management from the standpoint of accumulation of information resources (systemic involvement of information potential) and accelerate digitisation in all areas of regional and sectoral management:

- types and areas of obtaining and selling information resources;
- location of productions, organisation of settlements, production processes;
- public service systems based on the involvement of smart technologies (utilities, health care, education, etc.);
- environmental control and monitoring systems (smart technologies - assessments, examinations, etc.);
- geoinformation support for decision making;
- sphere of organisation of public services (information consultations, provision of documents, organisation of control, etc.);
- attraction and use of intellectual and labour potential of territories;
- virtualisation and telecommunication support, communication and communication systems;
- the latest information technologies for business incubation - acceleration and virtualisation;
- decision support; development, implementation, control over various types of policy (social, demographic, environmental, cluster, support of small and medium business, financial, innovation and investment, etc.);
- security, risk assessment, conflict zones, etc.

Also, a priority vector and an urgent task of economic science in today's Ukrainian economy realities, socio-economic reforms are the development of mechanisms for information support of decisions on the region's development, made by the legislature and the executive.

To date, the Ukrainian public administration has recorded the presence of more than 300 state registers for the provision of state registration services. Such a fact should not exist from the standpoint of duplication of information, criminal concealment of information and other systemic violations, especially property rights.

A systematic study of the regulatory framework of regional informatisation and the practice of regional governance in Ukraine indicate an insufficient informatisation and implementation of new, advanced information technologies and information support in the management decision-making system. This hinders the formation and development of regions and the state in general, preventing the formation of adequate management infrastructure.

Insufficient development and lack of support for the methods and approaches implemented in the information resources management system is a conflicting issue in the recognised quality of informatisation of regional management. In addition, the process of introducing advanced information technologies in administrative management is constrained by the weak level of organisation of information flows, collection, processing, storage and provision of information. Ways and methods of formalising information in the modern management system may not always be applicable in the regional management system of socio-economic processes. At the same time, a significant role in regional management is played by information exchange, which forms a specific information space in a particular geographical region, within which all business entities perform their information functions [5].

The involvement of cloud technologies, which have already gained worldwide recognition, is of particular importance for Ukrainian regional management. Priorities for attracting cloud technologies should be recognised: statistical and analytical support; e-government; communications of government, business and the population; remote provision of services; provision of educational services; cooperation with enterprises of the region and other areas.

The development of information management methodology and practice makes it possible to consider new independent areas of management related to information resource management, information processing, implementation and use of information technology in enterprises and organisations, recognition in the European information space and global markets of smart specialisation. The circulation of information between the elements of the system, the system as a whole and the environment is a prerequisite for effective, balanced, harmonious management in any field of activity. Thanks to information processes, given vectors of change, the system can carry out purposeful interaction with the external environment, coordinate the relationship of its components, direct their movement to a specific goal. The detailed data become management information if the consumer uses them in management processes directly before or after certain transformations. Thanks to management information systems, management data is transformed into management information, digitised and transparent, relevant to users' needs.

A separate area of information management development is the problem of providing managers with the necessary information for management decisions. The development of effective management decisions is based on competencies such as creativity, sociability, multiculturalism, communication, cooperation, critical thinking and in market conditions is primarily determined by objective, comprehensive, reliable information about existing processes.

One of the most important factors that should increase the efficiency of regional government in Ukraine is the introduction of information technology to change the nature of management work, strengthen the analytical and statistical component in the management decision-making process. Professional knowledge in regional management is marketing, project management, business administration, modelling and programming. At the same time, the following areas of involvement of advanced tools and technologies in practice should be considered priority and promising:

- development of fundamental information tools and information technologies (on platforms of cloud technologies, neural networks, geoinformation products, smart technologies);
- improvement and updating of operating systems;
- involvement of relevant programming languages;
- dissemination of technologies Data Mining, 3D printing, blockchain;
- attraction of information resources, automated programming, neurocomputations, client-server architecture, geoinformation and other database and knowledge management systems;
- dissemination and activation of telecommunication, multimedia and text systems, expert systems;
- improvement of security and cryptosecurity technologies;
- formation of introductory applied IT (provide development and implementation of standard technologies and procedures for processing information resources; definition of approaches and methodological support on standard procedures for accumulation, systematisation and processing of information by types and industries).

4 Creation of a specialised automated information system

The effectiveness of the regional policy of socio-economic development of the region depends on the timeliness, reliability, completeness and objectivity of the information received, the efficiency of its processing and generation in the system of management decisions. These processes constitute a system of information support of the management mechanism with current development priorities, methods, principles of management, which are based on the results of scientific and technical information development, methods and ways of accumulation and processing of information. It is necessary to create an information system of regional management, which would provide the process of managing the region with quality and timely information during management decisions.

One of the ways to overcome this conflict situation is the introduction of a specialised automated information system. In this case, choosing a model of such a system, which will provide the maximum and fastest effect at the lowest cost, becomes especially important. The system's primary purpose is to assess regional economic processes and prospects for their development in terms of economic growth. Creating an information system of regional information management is information support of regional management, business projects, the ability to inform investors about the investment potential of regional enterprises, information and analytical support, covering various aspects of regional reproduction processes. The results of economic analysis and forecasting can also form the basis for regional economic development.

The primary purpose of the development and implementation of the information system is information and analytical support of the management process, which helps reduce uncertainty and risks when making strategic and tactical decisions [6, 7, 8].

Requirements for information support of management activities in the regional management system also determine the unique, essential characteristics of information about the subjects of management to which it is intended. These features have their own theoretical and methodological basis, outline the formats of future development of this field of knowledge.

First, the information is intended for different management entities. Such information should be concise, diverse, relevant in content and as accessible as possible to a wide range of users.

Second, the information that enters the management system must be systematic, efficient, meaningful and objective. Due to limitations or bias, the essence of the management decision may be distorted; its effectiveness may be reduced after its adoption. In such cases, it is justified to use information from various sources to fully provide the authorities and eliminate the facts of failure or asymmetry in the information support.

Third, the information should be accompanied by targeted recommendations and relevant recommendations because, as practice shows, the management system always lacks unique and reference sources, integrated data (statistical collections, expert assessments, dictionaries, directories, geographic information resources, targeted innovative-applications).

The functional part is always related to the model of the management system of a specific object; in our case, it is the implementation of programs and projects of regional development that are relevant and meet the needs of the time.

We recommend that this area include issues related to:

- nature management and resource balance;
- formation of a system of resettlement, migration flows;
- evaluation and involvement of labour potential;
- development of interregional communications, intersectoral complexes and production infrastructure;
- targeted regulation of cluster formation processes;
- improvement of socio-cultural infrastructure;
- administrative-territorial development;
- systematic assessment of the development and support of small and medium-sized businesses;

- integration of government and business actions in the form of networks, technology parks, business incubators and accelerators, educational projects;
- targeted regulation of regional and sectoral markets.

We will give a brief description of the main elements of the information system of regional management, which are relevant in the formats of practical development and a further improvement in the knowledge system of regional information management, providing the involvement of knowledge management approaches [9-14].

Technical support - a set of technical means designed for the operation of the information system and relevant documentation for these means and technological processes.

Mathematical software is a set of mathematical methods, models, and algorithms for processing information used to solve problems and make management decisions. For instance, there is mathematic linear regression, which is widely spread in the use of researches, forecasting and development.

Statistical support is a set of statistical methods, models, and algorithms for processing information to create a management system.

The program software is a set of programs for implementing the goals and objectives of the information system, the normal functioning of a set of technical means.

Methodical and organisational support - a set of methods, tools and documents governing the interaction of information system personnel with technical means and with each other in the development and operation of the information system.

Legal support - a set of legal norms governing information systems' creation, legal status, and operation.

Information support of the management process is one of the functions of creating conditions for the effectiveness of management activities at all its levels: from an individual official to the highest public authorities. This function in the theory and practice of social systems management is called information and analytical support.

We recommend that the information support of the management process be effective to involve some requirements. Among them are:

- minimum information - provides for the restriction of collection and processing only those data (information), without which it is impossible to manage the object;
- optimal information - information meets the main requirements, comes high quality and on time.

Sufficiency of information - means that the information should give a comprehensive and complete picture of the operation of the object, as well as optimal or similar solutions. When determining the needs for certain types of information, we must first consider the importance of managing each type, the frequency of receipt, the relationship with other types of information.

The Ukrainian regional management system often lacks information because not all data included in the management system are helpful, presented in the proper form and meet the objectives of management. For example, some statistical reporting forms contain not so many news reports as reference and calculation indicators. There is no information on the state and quality characteristics of the business environment, cluster formation processes, the latest forms of business, renewal and modernisation of production.

As a result, there is a contradictory situation. When the management bodies are overloaded with generalised information on the economic development of the territory and economic agents, decisions are often made in conditions of uncertainty (conflict of information), which significantly reduces their quality.

One of the main requirements for the management process is to ensure the high reliability of the data used in decision-making processes. Without accurate information about the object of management, it is impossible to organise its work, use the most advanced management methods.

5 Conclusions

Providing the necessary information to specialists at all levels has a positive effect on creating a rational, harmonised organisational structure of management and allows the application of new methods, increased productivity, and the overall level of management culture.

One of the main requirements for the management process is to ensure the high reliability of the data used in decision-making processes. Without accurate information about the object of management, it is impossible to organise its work, use the most advanced management methods.

At the regional level, the main tasks of management information support should be:
- creation of new generation regional management information systems (based on digitalisation tools and taking into account the best world experience);
- identifying the needs of all stakeholders in new information technologies and databases;
- development of standard projects and standards of information support of the regional management system;
- improvement of statistical reporting in the system of regional management (according to the latest standards of business evaluation, social sphere, information component of development);
- preparation of normative-legal acts on the issues of information support of the administrative system of regional administration bodies, in particular the system of protection against information threats.

The solution of the set tasks is possible due to the creation of the subsystem of information-analytical support, which should be aimed at high-quality management of socio-economic processes to achieve stability, self-sufficiency, and ensuring normal living conditions of the region. The creation and further operation of the subsystem should be carried out taking into account modern requirements and technical and technological capabilities based on available means of processing, storage and provision of information, information technology and more.

The introduction of modern information technologies and improvement of informatisation of management processes can form a basis for growth of efficiency of administrative activity in the long run, strengthen competitive positions of regions, and provide a necessary level of information security. At the same time, one of the most critical factors of effective management is the use of information potential as a component of the socio-economic, information regional potential of Ukraine.

References

1. V. Bero, Yu. Chupailenko, Problems of Innovation and Investment Development, **5**, 9-16 (2013)
2. I. Shevchuk, *Information technologies in the regional economy: Theory and Practice of Implementation and Use,* Monograph. (Lviv, NNVK "ATB", 2018)
3. Virtual Museum of Canada (VMC). 30.03.2011, http://www.museevirtuel-virtualmuseum.ca
4. O. Zinchenko, *Formation of positive potential regions' image of Ukraine: Theory, Methodology, Practice,* Monograph (Cherkasy, R. Ponomarenko edition, 2018)
5. Yu. Kovalenko, Herald of Donetsk State University of Management, **35**, 33-38 (2018)
6. K. Beliakov, *Informatization in Ukraine: problems of organisational, legal and scientific support,* Monograph (Kyiv, KVIC, 2008)

7. Yu. Kovalenko, *Problems and features of regional information management development,* Monograph (Cherkasy, R. Ponomarenko edition, 2018)
8. O. Zakharova, N. Podluzhna, O. Yakusheva, and O. Yakushev, *Regional policy of advanced ICT in the minds of forming economy knowledge, in Proceedings of the 16th International Conference on ICT in Education, Research and Industrial Applications. Integration, Harmonisation and Knowledge Transfer*, ICTERI, Kharkiv, Ukraine, **2740**, 386-391 (2020)
9. I. Sotnyk, D. Hulak, O. Yakushev, O. Yakusheva, O.V. Prokopenko, and A. Yevdokymov, Polityka Energetyczna – Energy Policy Journal; **23(3)**, 147-164 (2020)
10. K. Wiig, *Knowledge Management 20 Years After. The Evolution and Increasing Significance of Knowledge Management,* http://www.krii.com/downloads/KM20 Years Later.pdf (2018)
11. Software and Information Technology Spotlight, https://www.selectusa.gov/software-and-information-technology-services-industry-united-states (2018)
12. C.P. Francisco, D.V.J. Nilson, Journal of Technology Management & Innovation. **3(4)**, 80-94 (2008)
13. M. Abbasi, P. Vassilopoulou, L. Stergioulas, Creative Industries Journal, **10(1)**,1-19 (2017)
14. C. Chapain, N. Clifton, and R.Comunian, City, Culture and Society, **(5)**, 51–53 (2014)

Facebook and Instagram Business Account Management under the Pandemic Influence

Alona Natorina*

Kyiv National University of Trade and Economics, Department of Trading Business and Logistics, 02156, Kyiv, 19, Kyoto str., Ukraine

Abstract. The crucial role of maintaining business accounts on social media for the rational organization of goods promotion in the COVID-19 pandemic context and increasing online business awareness is reasonably justified. Based on empirical research and analysis of Ukrainian retailers' online stores, business account metrics are diagnostics in social media. The interpretation of the viral loop in social media is proposed, and the meaning of the definitions "subcritical mass" and "critical mass" of a Facebook / Instagram business account are clarified. The dynamic model of a critical mass of Facebook / Instagram business accounts is developed. The model allows monitoring the nature and influence of active followers on users to adjust the current vector of activity, minimize marketing costs, and increase the target audience involvement. The model also facilitates accurate targeting and error-free ad placement for online businesses. Keywords: social media, business accounts, online business, Facebook / Instagram business account, account management.

1 Introduction

In the context of the globalization of the business environment, characterized by the scale of digital transformations and intensive competition, social media have a significant impact on developing an effective strategy for online businesses as they guarantee their progressive economic development. In particular, over the past ten years, the most popular social media in Ukraine – Facebook (Fig. 1), even though since 2017 it has had a steady decline in its market share. For example, the most popular social media among Ukrainian Internet users during 2011-2021 were Facebook, Twitter, Pinterest, Instagram, YouTube, Reddit, and Tumbler. Twitter and Pinterest were in second and third places in popularity among Ukrainians.

It is worth noting the significant increase in the audience of Twitter, whose market share increased in 2021 by about two times compared to 2020, due to the introduction of new and user-friendly innovations. Facebook, Pinterest, and YouTube at the beginning of 2021 lost some of their Internet users compared to 2020. Thus, Facebook's market share at the beginning of 2021 was 61.29 %, which is 7.86 % less than in 2020 and 24.85 % more than in 2017. This is due to the growing interest of Internet users of the social media Instagram.

* Corresponding author: a.natorina@knute.edu.ua

If in 2017 the market share of Instagram was 0.08 %, then by the beginning of 2021, its value increased by 8.45 %. In 2021, compared to 2020, the growth rate of Instagram's market was 287 %. In addition, it should be noted that taking into account the peculiarities of the content on Instagram, since 2019, YouTube began to lose a significant part of its audience, and its market share by the beginning of 2021 was, for the first time in 5 years, less than on Instagram and was 3,21 %.

Taking into account the above and interests of Internet users, online business should use the capabilities of popular social media to develop and promote their business effectively.

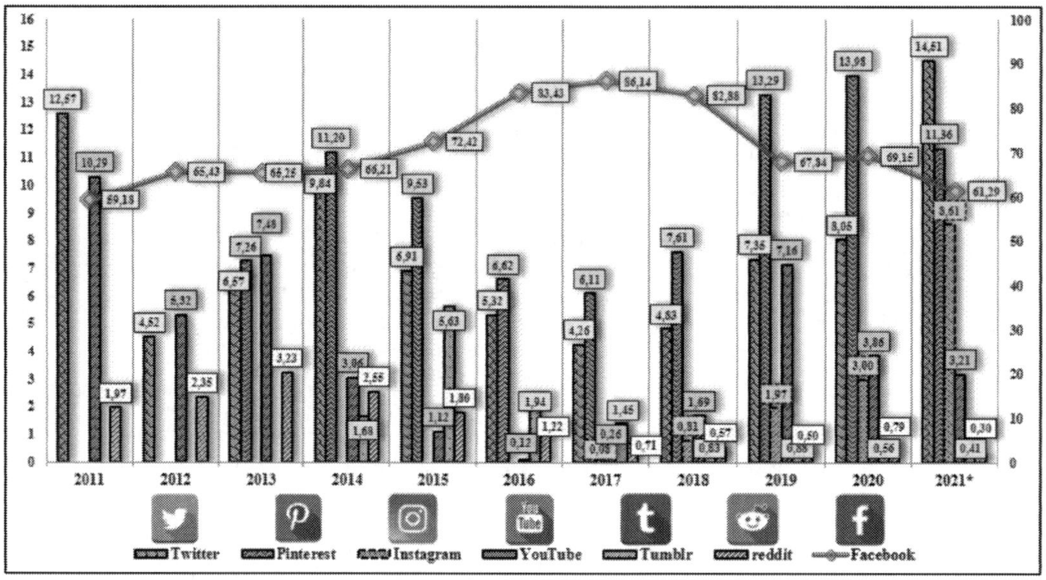

Notes. * – January-February, 2021

Fig. 1. Market shares of the most popular social media in Ukraine, % (calculated and created by the author according to [1])

2 Brief Literature Review

The positive impact of social media on online business development with relevant communication strategy that is studied and interpreted in works [2-6]. The specifics of business accounts administration in social media are described in publications [7-10]. The main advertising content characteristics in social media and the role of an online platform for attracting the target audience are thoroughly reflected in articles and reports [11-16]. Despite the significant scientists' contribution and without diminishing the importance of scientific achievements, it should be stated; it is necessary to mention the fragmentation of the study of business account management in social media.

3 Purpose

The study aims to develop methodological tools and guides for an online business that allows monitoring the effectiveness of Facebook / Instagram business account management.

4 Results

4.1 Empirical research background

As part of empirical study and complex analysis of the online business of Ukrainian retailers [17-18], the systematicity of updating business accounts content of retailers' online stores in social media is determined. Additionally, it should be noted that all retailers were divided into three clusters:

Cluster 1: Auchan Ukraine (FR1), Metro Cash and Carry Ukraine (FR2), NOVUS Ukraine (FR4), Yves Rocher Ukraine (DR1), RUSH (DR2), ALLO (HA1), Foxtrot (HA2), DIESA (HA3), Comfy-Trade (HA4), Harazh Mobail Hrup (HA5), Citrus Discount (HA6), Leroy Merlin Ukraine (DIY5).

Cluster 2: NASH KRAI (FR3), Tavria V (FR5), Fozzy Food (FR6), Budmax (DIY3).

Cluster 3: BRV Kyiv (DIY1), Nova Linia (DIY2), Epicentr K (DIY4), Mebelna Kompaniia Ukrainy (DIY6), JYSK Ukraine (DIY7).

Table 1 represents business accounts of Ukrainian retailers' online stores in social media and the regularity of their content updating.

Table 1. Business accounts of Ukrainian retailers' online stores in social media and regularity of their content updating (built by the author based on research in social media, January 2021)

Metrics	Facebook business account	Systematically updated content on Facebook	Instagram business account	Systematically updated content on Instagram
Cluster 1	FR1, FR2, FR4, DR1, DR2, HA1, HA2, HA3, HA4, HA5, HA6, DIY5	FR1, FR2, FR4, DR1, DR2, HA1, HA2, HA3, HA4, HA5, HA6, DIY5	FR1, FR2, FR4, DR1, DR2, HA1, HA2, HA3, HA4, HA5, HA6, DIY5	FR1, FR2, FR4, DR1, DR2, HA1, HA2, HA3, HA4, HA5, HA6, DIY5
% of retailers	100	100	100	100
Cluster 2	FR3, FR5, FR6, DIY3	FR3, FR5, FR6	FR3	FR3
% of retailers	100	75	25	25
Cluster 3	DIY1, DIY2, DIY4, DIY6, DIY7	DIY2, DIY4, DIY6, DIY7	DIY1, DIY2, DIY4, DIY6, DIY7	DIY2, DIY4, DIY6, DIY7
% of retailers	100	80	100	80

The data obtained during the empirical study (Table 1) show that the business accounts of Ukrainian retailers on Facebook and Instagram play a crucial role for most Ukrainian online retailers, in particular in building relationships with the target audience and increasing its level of loyalty. It is essential to mention that six retailers from the total quantity (FR1, FR2, FR4, FR3, FR6, DIY4) had Facebook and Instagram business accounts, which function separately from the online storefront.

4.2 A dynamic model of a critical mass of Facebook / Instagram business account

The author developed the dynamic model of critical mass of Facebook / Instagram business account to solve the above problem. The main advantage of this model is the systematic tracking of changes in the number of active followers and identification of the nature of their influence on social media users. Moreover, the proposed model is the foundation for making balanced management decisions for an online business to determine the current actions on the Internet. The critical mass of a Facebook / Instagram business account is an accelerator for an increase in the number of potential buyers and, consequently, a determinant of the multiplier effect from sales. The subcritical mass of a Facebook / Instagram business account is the number of followers, which is not enough to attract new followers and activate the viral loop. The minimum number of followers that form the subcritical mass depends on the specifics and scale of the business and is determined based on empirical research. Fig. 2 graphically interprets the subcritical (a) and critical (b) mass of a Facebook / Instagram business account.

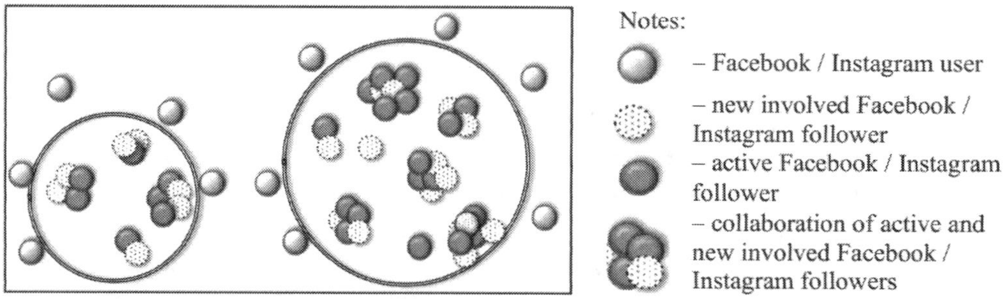

a) b)

Fig. 2. The subcritical (a) and critical (b) mass of a Facebook / Instagram business account (created by the author)

According to Fig. 2, the critical mass of a Facebook / Instagram business account depends on the interaction frequency, particularly of the active followers' communication with Facebook / Instagram users, who form the target audience. Therefore, with a low level of interaction between active followers of a business account and users who form the target audience on Facebook / Instagram, the rate of building up the critical mass – decreases. Thus, the guarantee of the rapid growth of the Facebook / Instagram business account critical mass is an increasing number of active followers to accelerate the passage of Facebook / Instagram users through the three stages of becoming an active follower and expanding the target audience, as well as encouraging it to make purchases in the online store.

Considering the above, online businesses should aim to intensively increase the number of active followers of the Facebook / Instagram business account by planning, developing, and implementing relevant marketing campaigns, considering the type and the target audience characteristics. In addition to traditional means of the target audience segmenting in social media, online businesses should use a parser. This program/script provides a particular mathematical model for its development and allows to scan, compare, and clearly identify the target audience preferences.

Nevertheless, in practice, it is impossible to accurately predict the time to reach the desired critical mass on the Internet, the speed and rate of its growth. However, systematic tracking of changes in the critical mass of a Facebook / Instagram business account will identify specific trends and make an informed online business management decision in the framework of following the general goal and implementing short-term and long-term measures. Based on this, the dynamic model of a critical mass of Facebook / Instagram business account was developed (1) in order to monitor and conduct a retrospective analysis of the total number of Facebook / Instagram business account followers, as well as to identify changes in the subcritical mass and the effectiveness of the influence of active followers on the conversion from Facebook / Instagram users to new followers. This will correct if it is necessary for online business, choose a vector of activities, and implement a plan to increase the critical mass of a Facebook / Instagram business account in perspective.

$$M_{cr}=M_{sub}+F_n-F_{un}, \tag{1}$$

Where M_{cr} – critical mass; M_{sub} – number of followers that form a subcritical mass; F_n – number of new involved followers; F_{un} – number of unsubscribed followers. So, F_{un} can be identified as the loss of online business share of the target audience in Facebook / Instagram because of a new competitor and the transition of followers to it.

According to (1), when $F_n > F_{un}$, then M_{cr} increases, and $M_{cr} \to \infty$. Otherwise, when $F_n < F_{un}$, then M_{cr} decreases and $M_{cr} \to 0$. With a sharp decrease in the number of followers, it is possible a situation when $F_n = F_{un} = 0$, a $M_{cr} = M_{sub}$.

It is important to emphasize that F_n reflects the cumulative amount of new Facebook / Instagram business account followers, attracted by active followers and those Facebook / Instagram users who became new business account followers, thanks to the successful marketing policy implementation (2):

$$F_n = F_{inv} + F_{ind}, \tag{2}$$

Where F_{ind} – number of users who became new followers without the influence of active followers on them (in the absence of interaction with active followers); F_{inv} – number of new followers attracted by active followers. Thus, the dynamic model of a critical mass of Facebook / Instagram business accounts is (3):

$$M_{cr} = M_{sub} + F_{ind} + F_{inv} - F_{un} \tag{3}$$

The average number of new followers attracted by an active Facebook / Instagram follower of a business account (F_{inv_av}) (4):

$$F_{inv\_av} = \frac{F_{inv}}{F_a}, \tag{4}$$

Where F_a – number of active Facebook / Instagram business account followers.

According to (4), online businesses should strive to achieve the maximum average number of new attracted followers Finv_av=Finv/Fa under Finv→max and Fa→min conditions. At the same time, the time *t,* which spends an active follower on communication with a Facebook / Instagram user to turn him into a new follower, plays a significant role in achieving Finv_max. Therefore, $F_{inv\_av} \to max$, $F_{inv} \to max$, $F_a \to min$, $t \to min$. The visualization of the dependence of M_{cr} on t and F_a in the dynamic model of a critical mass of a Facebook / Instagram business account is represented in Fig. 3.

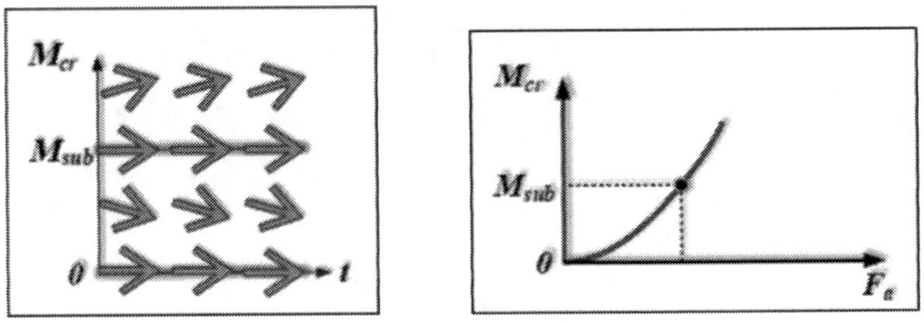

a) b)

Fig. 3. Dependence of Mcr on t (a) and Fa (b) (developed by the author)

According to Fig. 2, if $M_{cr}=0$, then the solution of the model is stable, while $M_{cr}=M_s$ – unstable. The last one is due to the inability to predict Fn and Fun accurately. With $M_{cr}=M_{sub}$, one can observe both an increase and a decrease in the number of followers, which depends on the online business marketing policy and the chosen vector of activity. When Mcr>Msub indicates a continuous increase in Fa, and $M_{cr}<M_{sub}$ – a constant decrease in F_a.

5 Conclusion

Using the developed dynamic model of a critical mass of a Facebook / Instagram business account in practice will make it possible to systematically track changes in the number of followers and show the nature and influence level of active followers on Facebook / Instagram users. Also, this model will serve as the basis for making informed management decisions to change the vector of online business activity in social media to reduce the cost of further marketing activities implementation.

References

1. Statcounter GlobalStats (2021), URL: https://gs.statcounter.com/
2. G. Appel, L. Grewal, R. Hadi, A. Stephen, Journal of the Academy of Marketing Science, **48**, 79-95 (2019), URL: https://link.springer.com/content/pdf/10.1007%2Fs11747-019-00695-1.pdf
3. Capgemin (2021), URL: https://www.capgemini.com/
4. CISCO (2021), URL: https://www.cisco.com/c/uk_ua/index.html
5. J. Plantin, C. Lagoze, P. Edwards, C., New Media & Society, **20 (1)**, 293-310 (2018)
6. S. Yermak, O. Bavyko, O. Boiko, N. Vasiutkina, Journal of Hygienic Engineering and Design, **27**, 15-21. (2019), URL: http://www.jhed.mk/filemanager/JHED%20Vol.%2027/01.%20Full%20paper%20-%20Svitlana%20Yermak.pdf
7. CIO (2021), URL: https://www.cio.com/
8. H. B. Dondolo, Problems and perspectives in management, **15 (2)**, 404-410. (2017)
9. R. G. Duffett, Young Consumers, **18 (1)**, 19-39 (2016), DOI: https://doi.org/10.1108/YC-07-2016-0062
10. M. Bondarchuk, O. Bavyko, Modern science – Moderni věda, **3**, 5-13 (2019)
11. Content Strategy Alliance, *Content strategy alliance tools and templates. A best practices handbook* (2015), URL: http://contentstrategyalliance.com/wp-content/uploads/2015/08/CSA-Handbook.pdf
12. EVO.business (2021), URL: https://evo.company/ua/
13. Isobar Ukraine (2021), URL: https://isobar.com.ua/
14. D. Lee, K. Hosanagar, H. Nair, Management Science, **64 (11)**, 1-41. (2018), DOI: http://dx.doi.org/10.1287/mnsc.2017.2902
15. Nielsen (2021), URL: http://www.nielsen.com/
16. H. Virtanen, P. Bjork, E. Sjostrom, Journal of Small Business and Enterprise Development, **24 (3)**, 468-484. (2017), DOI: https://doi.org/10.1108/JSBED-12-2016-0202
17. A. Natorina, Baltic Journal of Economic Studies, **5 (1)**, 131-136, DOI: https://doi.org/10.30525/2256-0742/2019-5-1-131-136 URL: http://www.baltijapublishing.lv/index.php/issue/article/view/610
18. A. Natorina, Economic Annals-XXI, **174 (11-12)**, 69-72 (2018), DOI: https://doi.org/10.21003/ea.V174-11

Main Points of Digital Agency Business Processes Management: Foodtech Sector Case

Elena Omelyanenko[1*], *Teofilo* Tirto[2], *Dmitry* Volodin[2], *Vitaliy* Omelyanenko[1,3], and *Galina* Kovtun[1]

[1]Department of Business Economics and Administration, Sumy State Pedagogical University named after A. S. Makarenka, 40002, Romenska str., 87, Sumy, Ukraine
[2]R&D Department, FARADI SRL, Italy
[3]Teadmus OÜ, Estonia

Abstract. The research deals with the analytics of the features of business processes of a digital agency and the identification of promising aspects of digital marketing as a tool for B2B interaction in the Foodtech market. It is noted that the primary trend of the Foodtech sector deals with digital transformation. From the strategic point of view of creating a business model, it is necessary to understand that digital transformation involves the installation of modern hardware or software and fundamental changes in approaches to management, corporate culture, and communications. To increase the efficiency of the digital agency's business model, the authors propose to consider digital marketing. The analysis of the effectiveness of the business model of the digital agency allowed us to identify the most problematic and, at the same time, promising components of the business model. Based on generalized approaches to business model management in the case of a digital agency in the design of relevant services, it is proposed to consider maximizing the speed and efficiency of business model transformation and minimizing the uncertainty of further digital service development based on customer relationship management. Keywords: Foodtech sector, digital agency, business processes management, digital transformation.

1 Introduction

The analysis of the practical implementation of digital marketing allowed us to conclude that in rapid digital transformations, active market participants are digital agencies that provide a wide range of services to solve creative and technical problems for various services in the Internet space.

Digital agencies play a unique role in the implementation of digital marketing functions because, in contrast to media or marketing agencies that provide platforms for advertising, offer information and analytical services, which can determine the specifics of business models in the digital sphere:

– standard services (site creation, media and contextual promotion, design, site

* Corresponding author: elnikkrasnaya@gmail.com

development);

– development of a comprehensive strategy for the development of the enterprise in the digital environment (examination and promotion);

– work with online communities (groups and pages in social networks, blogs, forums, specialized sites);

– organization of event events, competitions (strategy, creative, implementation, performance analysis) in combinations of online/offline promotion;

– consumer transition from online to offline and back (promotions, etc.);

– experimental marketing, QR codes, geolocation services, RFD, WOW-calls, augmented reality, etc.

Based on the considered range of digital agency services, the primary task of digital marketing is to identify and attract both traffic (for customers) and potential customers (for the customer). This can be done by reviewing organizational capabilities (strategy, culture, change management, innovation management and customer path maps).

In our study, we will analyze the features of the business processes of a digital agency and identify promising aspects of the use of elements of digital marketing as a tool for B2B interaction in the Foodtech market.

2 Foodtech sector business processes management trends

The primary trend of the Foodtech sector deals with digital transformation. From the strategic point of view of creating a business model, it is necessary to understand that digital transformation involves the installation of modern hardware or software and fundamental changes in approaches to management, corporate culture, and communications. As a result, productivity and customer satisfaction increase, and the company acquires a reputation for being progressive and modern.

Digitalization of business processes is relevant not only at the level of individual enterprises. In current conditions, some industries choose this development path as the only way to meet the dynamic market conditions. Not inferior to the above sectors and the restaurant business, the integration of digital technologies already forms a whole segment – Foodtech. In the foreign market, we can identify the following main areas of Foodtech:

– delivery of food from restaurants and cafes through online arrangements;

– delivery of food through online arrangements;

– Possibility to pre-order meals up to a specific time / take away food;

– availability of a large number of services with a broad base of restaurants and the ability to order meals online.

As a result, the digitalization of the restaurant business model takes place in two directions: communication with the customer goes online, and the large-scale introduction of automation reduces the number of staff involved in service and management.

According to the study, one of the stable trends in recent years is the growing popularity of loyalty programs. Special offers and marketing promotions are now available in almost all restaurants, and customers before visiting the institution, customers at least browse the institution's site, the public on social networks, and read reviews. In addition, in the face of growing competition, the classic bonus programs are being replaced by more advanced promotions, programs with elements of gamification, etc.

Loyalty programs are also digitally converting en masse. Instead of plastic cards, mobile applications combine a bonus system, the ability to book tables, information systems, special offers, personal recommendations, food diaries, and feedback tools.

In contrast to the foreign restaurant market, where digitalization has now reached a high level (the share of penetration of online service in the food segment in the US is 3%, in the UK 7%), Foodtech in Ukraine is just beginning to be implemented – the share of online services is no more a few per cent. According to analysts, the restaurant business in 2021 shows some growth and is actively developing the field of e-commerce.

These factors indicate the need for active implementation and development of Foodtech because the catering market is waiting for a change shortly. From the practical point of view of Foodtech implementation, the restaurant business's main task will be to make ordering as easy as possible and innovate to make ordering possible in one or two clicks.

It is necessary to analyze the market and the target audience's demands to solve this problem. Current trends in the e-commerce market and online orders dictate the need to use advanced analytics about the target audience.

As a rule, restaurants do not have such statistics and, accordingly, do not have flexible and convenient tools for proactive management of the business model, thus sales, promotion, and revenue growth.

It is necessary to have complete information about the target consumer and to personalize the data about him as much as possible to build a business model and provide quality service. This will allow us to effectively use advertising channels and regulate workflows in the right direction.

In these conditions, implementing the system (system elements), Big Data allows you to get advanced analytics for users in any field, which opens vast opportunities for business owners. Ability to personalize, improve the quality of services, change and expand the range can significantly improve the quality of service or service. With the development of e-commerce and the Foodtech market, advanced analytics will become increasingly necessary in business.

Gartner experts believe that the success of digitalization depends not so much on modern technology as on the availability of built production processes and systems. Skills, initiatives and business models are 15 times more important for success in digitalization than disparately implemented technologies [mosto].

The business intelligence system helps reduce costs through its optimization. You can accumulate all the data on one platform and make decisions based on the real situation in real-time. In other words, automation of analytics allows us to manage the enterprise most effectively.

From this point of view, the main task of the restaurant management system is to manage information flows, and business intelligence systems allow us to store and process large amounts of data.

Break-even investment in the business process management system arises with the beginning of a systematic analysis of the business processes of the enterprise in order to strategically improve them. Therefore, developing a model of analytical indicators of business processes is one of the most critical and complex tasks of business intelligence development. According to the author of the study [1], levels of decomposition have the following purpose in business intelligence:

– strategic level "Activity" is an abstract model of planning the goal vector and determining the priorities of business system development;

– tactical level "Processes" (business functions) is designed to ensure the balance of the system based on functional budgets in the development process, as well as for timely response to critical changes in the external and internal environment of the enterprise and adaptation to them;

– the level of "Operation" involves the actual management based on the strategic objective function and tactical budget constraints, measured and improved parameters of specific actions.

Thus, the preliminary analysis of business processes in the introduction of digital technologies in the restaurant business, in addition to potential conveniences for customers, opens up many new opportunities for the business itself. Among them, we can note the optimization of the kitchen and the ability to plan loading, revenue growth by reaching a new audience. In these conditions, the services and platforms of specialized business analytics allow us to bring to a new progressive level of customer service and effective marketing management.

Note that previously to enter the online market successfully, the entrepreneur had to order a specific site, create several pages (landing page) using one of the online services, buy a subscription base (at best – to collect), give a few advertising posts in social networks and understand the basics of SEO promotion. This provided a sufficient level of profit. However, in current conditions, everything is changing, and Internet technologies for business are not enough, traditional advertising only spends the budget, people's interest quickly fades, one wrong decision – and the subscriber base is already ineffective. At the same time, new digital communication channels appear with a wide range of opportunities for promotion, which leads to a change in the business model and forms the field of digital marketing.

3 Equations and mathematics

To increase the efficiency of the digital agency's business model, we propose to consider digital marketing. Digital marketing as an element of strategic management in building a business model can be defined by the following theses:

– a comprehensive approach to the promotion of the enterprise, its products and services in the digital environment, which also includes offline consumers using games, mobile phones and other digital means of communication;

– integration of a large number of different technologies (social, mobile, web, CRM-systems, etc.) with sales and customer service;

– ensuring constant quality two-way communication between the advertiser and the end-user of the product (service);

– the ability to combine technology and human resources, maintaining the right balance based on the needs of the target audience and the properties of the proposed product;

– phenomenal dynamics of growth and achievement of any marketing goals;

– the ability to be relevant to market requirements, evaluate and analyze the promotion results, flexibly respond to customer needs and adjust our product (service).

The analysis of the digital agency business model effectiveness allowed to identify the most problematic and at the same time promising components of the business model (Table 1).

Table 1. Components of business model with negative status.

Component of business model	Factors	Weight	Current status	Rating
Customer relations	Assistance in improving business processes	22,73	-2	-45,5
	Support of projects after their implementation	22,73	-1	-22,7
Sales channels	Use of digital marketing technologies	50,00	-1	-50,0
Key partners	Existence of strategic partnerships (B2B)	68,18	-1	-68,2
Key resources	Strategic management	86,36	-1	-86,4
Consumer segments	Consumer diversification	50,00	-1	-50,0
Costs structure	Investments	100,00	-1	-100,0

Based on generalized approaches to business model management in the case of a digital agency in the design of relevant services, we propose to consider maximizing the speed and efficiency of business model transformation and minimizing the uncertainty of further digital service development based on customer relationship management (in our case, restaurants and customers).

In traditional business models, consumers are also given importance, but in the case of a digital agency, it is necessary to create a system that meets the needs of both restaurant consumers and restaurant customers. Note that the inclusion of two groups of consumers in the service development process is traditionally limited to the initial and final phases of the project.

Note that from the standpoint of the M2 model, the primary purpose of the enterprise working in the field of digital promotion of restaurants, we can define as follows:

– ensuring interaction between project developers and the customer;

– observation and control of works, estimation of intermediate variants of services on the observance of requirements, correct performance, estimation of service and actual expenses;

– coordination of intermediate results of digital service design with the customer;

– checking the correctness of the digital service by testing it on planned and agreed with the customer test suites;

– assessment of compliance of digital service quality characteristics with the set requirements;

– discussion of the components of the digital service used in terms of assessing their capabilities and shortcomings identified in their application, as well as identifying areas for improvement or modernization of the service.

4 Conclusion

Thus, a digital agency is an enterprise that can provide restaurants (customers) with expert judgment and ensure the implementation of ideas in website creation, creativity, strategy, advertising and customer service. Digital agencies are not limited to creating websites. They work comprehensively to create and strengthen its name in the market and its positioning in the media using effective viral advertising.

From the point of view of creating a business model, the agency's work must not stop at the end of the project. Representatives of such an enterprise can continue to develop the project for years, promote it in the media, striving to maximize its image and customer loyalty.

References

1. D. Mogilko, Business Process Analytics. URL: https://www.businessstudio.ru/articles/article/analitika_biznes_protsessov/ (2018). 126
2. M. Mosto, *Digital doubles, open new opportunities for the instrument making and electronics industries*. URL: https://controlengrussia.com/innovatsii/novye-vozmozhnosti/ (2019)
3. Yu. Robul, I. Lytovchenko, Li Tchon, Ye. Nagornyi, O. Omelianenko, International Journal of Scientific & Technology Research, **9**, 4 (2020)
4. V.A. Omelyanenko, O.M. Omelianenko, *Digital component of relationship marketing in services sector, in Proceedings of the International Scientific Conference Marketing of innovations. Innovations in marketing*, December, 2020. Bielsko-Biala (2020)
5. V. O. Braslavska, N. Biloshkurska, M. Biloshkurskyi, N. Kliasen, O. Omelyanenko, International Journal of Computer Science and Network Security, **21**, 9 (2021)

Problems Caused by the Impact of COVID-19 on Small and Medium Enterprises Management and its Solutions

Badri Gechbaia[1*], *Amiran* Tsilosani[1], *Olha* Prokopenko[2,3], *Zurab* Mushkudiani[4], and *Ketevan* Goletiani[4]

[1]Batumi Shota Rustaveli State University, Ninoshvili/Rustaveli str. 35/32, 6010 Batumi, Georgia
[2]Estonian Entrepreneurship University of Applied Sciences; 10a Suur-Sõjamäe, Tallinn, Estonia
[3]Teadmus OÜ, Tallinn, Estonia
[4]Batumi Navigation Teaching University, Tamar Mefe Ave. 38, 0105 Batumi, Georgia

Abstract. The present paper has been made a comparative analysis of Small and Medium-sized Enterprises to determine the impact of the COVID-19 pandemic-related restrictions and lockdown on their activity. At the same time has been analysed pre- and post-pandemic periods. Also, the paper provides opinions about the ways how to mitigate and avoid impact on small and medium enterprises (SMEs) caused by pandemics, which might reoccur in the future. The importance of SMEs in the world and Georgian economy has been estimated in terms of added value, GDP, turnover, manufacturing products, and working place formation. It has been concluded that SMEs can handle some minor and short-term issues by themselves with the elaboration of the business continuity plan and crisis management scenarios, but these still are needed the active involvement of state aid. The latter implies financial support and professional, informational, and technical support as well, such as state internet-platform available for SMEs for interactive coordination and communication between the parties and for rapid action from the state on those fields which has been damaged mostly. Keywords: Small and Medium-sized Enterprises, comparative analysis, COVID-19, pandemic-related restrictions, enterprises management.

1 Introduction

Amid the worsening economic crisis in the context of the negative impact of the COVID-19 pandemic and the imposition of coercive restrictions, governments worldwide are taking unprecedented steps to support economic and business activity and preserve workplaces. COVID-19 became a serious challenge and a threat to Georgia's world economy and economy and other countries as well. The virus has become a global pandemic, which has harmed the world economy; the degree of its ultimate impact remains unknown and

* Corresponding author: gechbaia.badri@bsu.edu.ge

depends on the ratio of its spreading, the effectiveness of crisis management, social responsibility, and the distribution and effectiveness of the vaccine [1].

As mentioned above, the COVID-19 pandemic hit the world economy; the main subjects are small and medium-sized enterprises (from now on SMEs) and not corporations. The SMEs account for 90% of the total number of companies worldwide, provide 70% of jobs and generate 50% of world GDP.

The importance of SMEs in the global economy is highly recognised, as evidenced by the UN General Assembly Resolution of 2017, which declared June 27 as Micro, Small, and Medium-sized Enterprises Day to raise public awareness of the contribution of SMEs to global economics [2]. The UN pointed out that SMEs may be very resilient to changes in the world, although this makes them highly vulnerable to adverse economic impacts, primarily due to insufficient financial sustainability [3].

The relevance of the research is evidenced by the COVID-19 pandemic-related situation in the world over the past year, which has essentially changed the world economic situation and put humanity and the world's developed countries and their economies in a whole new reality [4]. Especially it is relevant to SMEs, which have faced significant challenges due to the current situation and the various types of restrictions.

This research aims to describe the economic situation caused by the COVID-19 pandemic in SMEs in the world and Georgia, to review the current situation, ongoing processes, and identify the problems that can be avoided or mitigated with state support and on their own.

This research aims to study the situation of SMEs in the world and Georgia, identify and find ways to solve the existing problems, analyse the best world practices, and prevent, mitigate, or minimise potential risks.

2 Research methods

In the process of research has been used methods of analysis, synthesis, and comparison. To achieve the goals have been used complex methods of information obtaining. Bibliographic studies were used to extract, sort, and analyse secondary data. The theoretical part of the research is the existing literature on the topic, particularly scientific articles, research, publications, data obtained from internet sources.

Statistical calculations were carried out during the study based on official sources (Eurostat, Geostat, and others). The primary materials necessary for publication were obtained from the National Statistical Office and various agencies.

3 Literature review

The differences in the criteria for defining SMEs in different countries are due to objective institutional features, differences in the economy's structure, and the goals of state policy, but anyway, they are based on the turnover, capitalisation, or the number of employees. For instance, in the EU, enterprises with less than 250 employees are medium-sized enterprises, with fewer than 50 employees - small, and less than ten employees - micro-enterprises (Figure 1). Additional criteria are the profit of the companies and their book value [4].

	Large Enterprises (≥250 Emplyees)	Medium-sized enterprises (50-249 employees)	Small enterprises (1-50 employees)
Number of enterprises	0.2	0.9	98.9
Number of employees	33.1	16.9	50.0
Added value of products	44.1	18.6	37.3

Fig 1. Distribution of companies in the non-financial sector of the economy as a percentage of the total number by category [7].

In Georgia, small and medium-sized enterprises are defined based on the average annual number of employees or annual turnover. From 2017 up to the present, medium-sized enterprises include all enterprises with an average annual number of employees from 50 to 249 and an average annual turnover of 12 million to 60 million Georgian Lari (GEL). Small enterprises with an average annual number of employees do not exceed 49, and the average annual turnover exceeds 12 million GEL [9].

Numerous assessments at the global, regional, and national levels point out the importance of SMEs in the economy, considering their share in the total number of enterprises, number of workplaces, as well as their contribution to the formation of GDP and added value.

According to the UN data, the category of micro, small and medium-sized enterprises includes about 90% of all enterprises globally, employing about 70% of the workforce and accounting for 50% of world GDP [2].

The total number of SMEs reached 22.2 million and accounted for 56% of the total added value in the non-financial sector (3.5 billion euros). The total number of employees in SMEs has reached 83.9 million, which is 67% of the total number of employees from where almost half are employed in only three areas of economic activity: Retail (20.7 million employees, 27% of small enterprises and 19% of medium-sized enterprises), manufacturing (15.8 million people, 14% and 33%, respectively), and construction (10.2 million people, 14% and 7% respectively). According to the 2020 report of the National Statistics Office of Georgia, the added value generated by SMEs in 2019 amounted to 59.3% of the total volume, and the number of employees was 65% [11].

The bi-year statistics report "the EU in the world" declares that labour productivity tends to be higher in large corporations. Thus SMEs tend to have relatively high employment and low-value indexes at the country scale [13].

The lowest employment rates in large industrial enterprises (respectively, higher employment rates in SMEs) during this period were observed in South Korea and Turkey, reaching about 25% and 33%, respectively. The employment rate in large industrial enterprises in Japan, Australia, and the UK is close to the average for the 27 EU member countries and does not exceed 50% [5]. In other countries, these indexes are significantly higher and reach 50% in Brazil (2014), 52.1% in Canada (2016), and 65.2% (2015) in the United States. The largest share of employment at large industrial enterprises was recorded in 2017 in Russia, where it reached 85.2% (Figure 2).

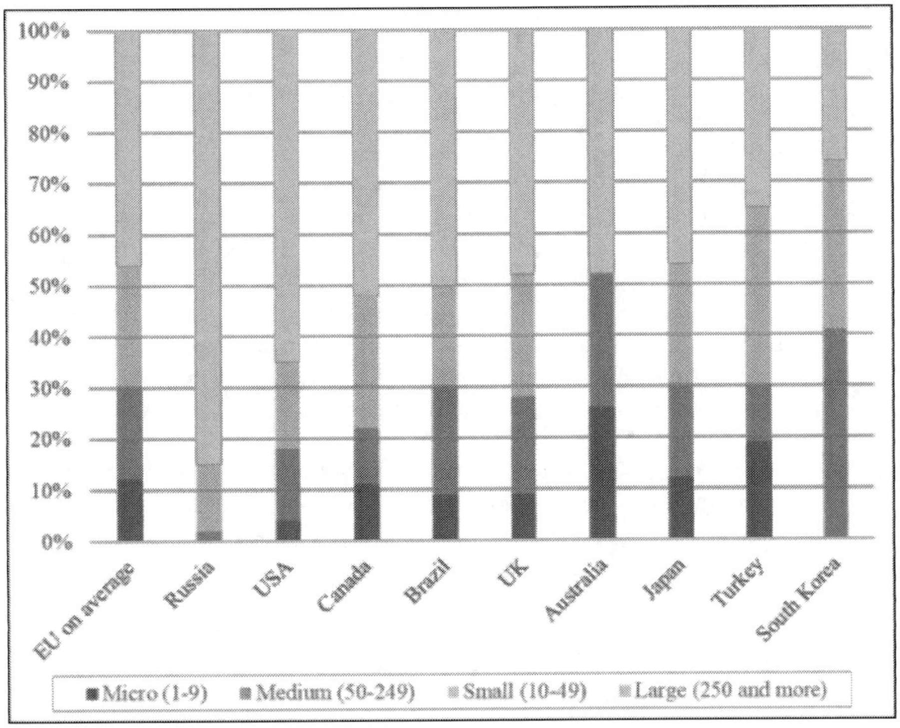

Fig 2. Distribution of industrial companies by workplaces in the percentage of the total number in the industry (2017) [8].

Despite the significant share of SMEs in the structure of the national economy in terms of the number of enterprises, workplace generation, and added value, their participation in foreign trade is less noticeable.

According to the European Commission, most European companies find it challenging to cross national borders even in a single market and equal trade terms. Only 25% of SMEs European businesses are involved in export activities, and even fewer of them - around 600 thousand - go outside the EU [8].

In addition, the share of exports in the turnover of such companies is small - only 7.3% of SMEs had a share of exports more than 10% of the trade turnover, considering that the share of SMEs in the total number of external economic entities of the EU is quite large, e.g., in 2012-2017 accounted for 97.8% of the total number of exporters [8].

Similar statistics are presented for American companies. According to the US International Trade Commission report in 2016, approximately 400,000 SMEs participated in foreign trade, accounting for 98% of exporters, accounting for 33% of the total value of exported goods. According to the US Small Business Administration, the number of SME-exporters in 2018 did not exceed 288 thousand, about 1% of the total number [10].

4 Results

SMEs are highly vulnerable to the adverse economic conditions caused by the COVID-19 pandemic. As indicated by the International Monetary Fund, this is associated with a significant reduction in market supply. Most companies are cutting production due to quarantine and other restrictions and material supply disruptions or labour shortages as employees [11].

At the same time, a sudden and significant decrease in the demand for goods and services leads to a decrease in income, which prevents SMEs from continuing to operate due to a lack of funds. The main reason is the change in consumer habits and behaviour due to fear of infection and a decrease in personal income and wages or layoffs, which ultimately lead to a circular effect. The effect of such social distancing has an especially harmful impact on SMEs [12, 13].

In addition, the statistical review of the Organization for Economic Cooperation and Development (OECD) emphasises that the share of SMEs is highest in those sectors that usually experience more economic problems in the current situation, such as engineering, construction, wholesale and retail, trade, hospitality, catering, real estate, and customer service. The abovementioned areas account for 40% of workplaces in the member countries of the Organization for Economic Cooperation and Development (OECD), with an average of 75% account for SMEs. For reference, this index reaches almost 90% in Greece and Italy [10].

A qualitative analysis of information on the impact of the pandemic on SMEs is possible based on regular surveys conducted by associations, international organisations, research centres, chambers of commerce and banks, e.g. one section of the report of Organization for Economic Co-operation and Development (OECD) contains 40 similar surveys/questionnaires, from which follows that most representatives of SMEs are experiencing the growing negative impact of the coronavirus pandemic and apprehend of closure of their businesses in the coming months [10].

In April 2020, big four audit company "PricewaterhouseCoopers" surveyed 1,938 companies to determine the impact of the spread of the coronavirus on micro, small, medium and large enterprises in Georgia [14]. Also, 33% of the companies surveyed were from the regions of Georgia. Also, 46% of the respondents were micro-enterprises, and 25% were medium-sized enterprises. The study showed that decreased demand in most businesses (about 65%) has already resulted in a loss of more than 50% of revenue, especially suffered hotel and restaurant businesses, and over 75% experiencing significant revenue reduction. By the survey has been identified that 28% of respondents already faced the problem of fulfilling their obligations, and more than 50% of respondents have already made staff cut-offs (the most significant reductions were made in the hotel and restaurant sector, where more than 40% of enterprises laid off staff fully) [9].

Despite the relatively favourable financial conditions in the pre-pandemic period for SMEs, they were insufficient to endure the economic outcomes on their own without government intervention and participation. They were especially funding difficulties expected for the companies with high risks, such as innovative enterprises and start-ups. Trade and supply chain disruptions, reduced revenues, increased unforeseen expenses will result in a liquidity crisis, which requires an additional audit [15]. Consequently, the steps taken by governments are aimed at supporting companies to cope with the severe but temporary liquidity shortages caused by the coronavirus infection.

The World Trade Organization agrees with the assessment that limited financial resources and borrowing capacity are the most significant factors in the vulnerability of SMEs in terms of the pandemic. According to a press release prepared by the WTO's secretariat, many SMEs depend on international trade to sustain their activities, whether they export their goods themselves or through intermediaries or import materials and composite goods into the domestic market for other sales. Moreover, such enterprises dominate precisely those sectors most affected by social distancing measures and transport restrictions [16].

With the advent of COVID-19, publications began to be published describing the harmful effects of the crisis on SMEs and analysing the relief measures and measures taken by the governments of different countries [17-18].

Supporting activities may significantly impact the state budget, the level of internal and external debt. In order to ensure proper governance, it is essential to ensure the transparency and independence of the parties when providing focused support to companies. All government decisions and statements must be cost valued, documented, and monitored. Should be set ways of monitoring and control of the actual risks to guarantees and other commitments. Also, should be considered the measures and ways of risk mitigation.

In December of 2020, the government of Georgia has adopted a series of fiscal measures to mitigate the economic impacts caused by the pandemic. The total amount of declared measures mentioned above was almost GEL 1.1 milliard (2% of GDP). It included GEL 545 support of citizens and GEL 515 for business support. The latter has been provided supporting activities such as partial withholding tax exemptions on payroll, full property tax exemptions, and subsidy of loans for companies related to tourism, small grants for entrepreneurs, defer bank loans, etc. [9].

The first lockdown in Georgia affected the fiscal outcomes of the country: on June 24, 2020, parliament approved an amendment to the state budget, which included measures of mitigation of the government's Anti-Crisis Economic Action Plan. The supplementary budget increased the budget deficit to 8.5% of GDP.

The Georgian economy faced several problems, such as the depreciation of the national currency, a sharp decline in activity in tourism and restaurants and related services, deterioration in import and export indexes (Figure 3), and reduction in foreign direct investment (Figure 4). It should be mentioned that some figures get better after lifting strict restrictions related to the COVID-19 pandemic and implementing fiscal measures by the National Bank of Georgia increasing of refinancing rate has been strengthened the national currency against the US dollar [19]. Also has been increased export YTD May export volumes as compared to 2020 and even pre-pandemic 2019 period. Worth mentioning that it also increased money transfers from abroad (Figure 5).

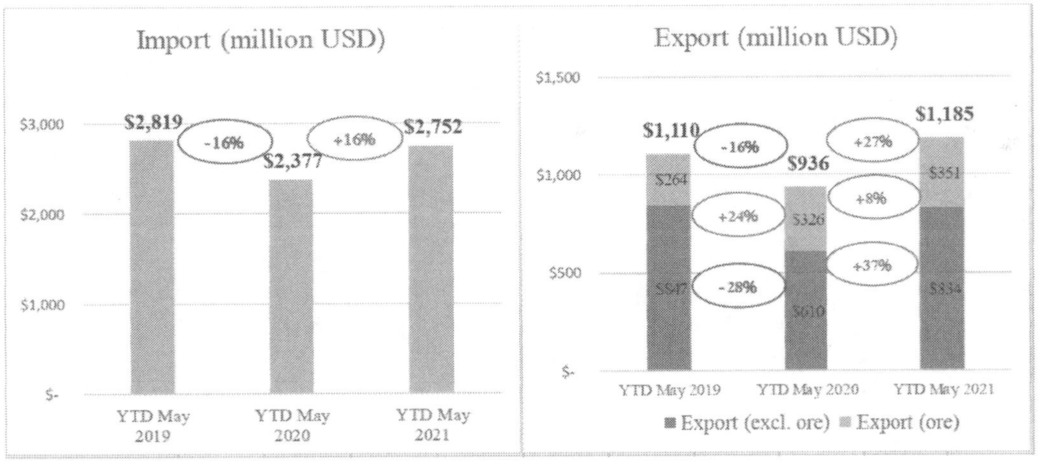

Fig 3. Import/Export figures for 2019 vs 2020 vs 2021 YTD May period [19].

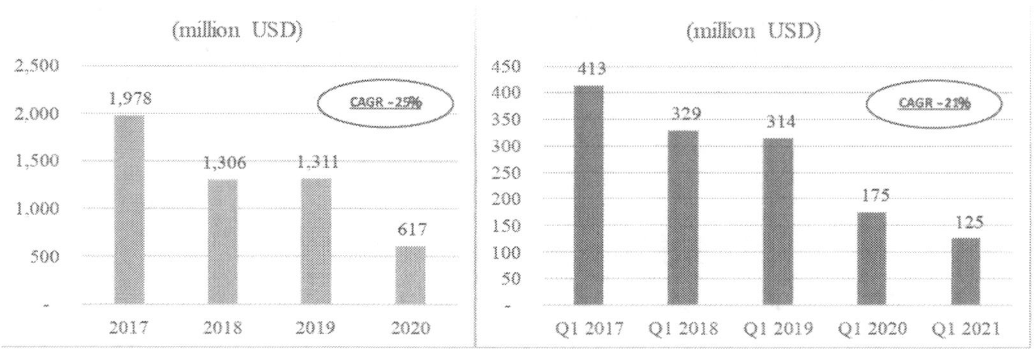

Fig 4. Foreign direct investment (FDI) whole period of 2017-2020 and 1 Quarter period of 2017-2021 [20].

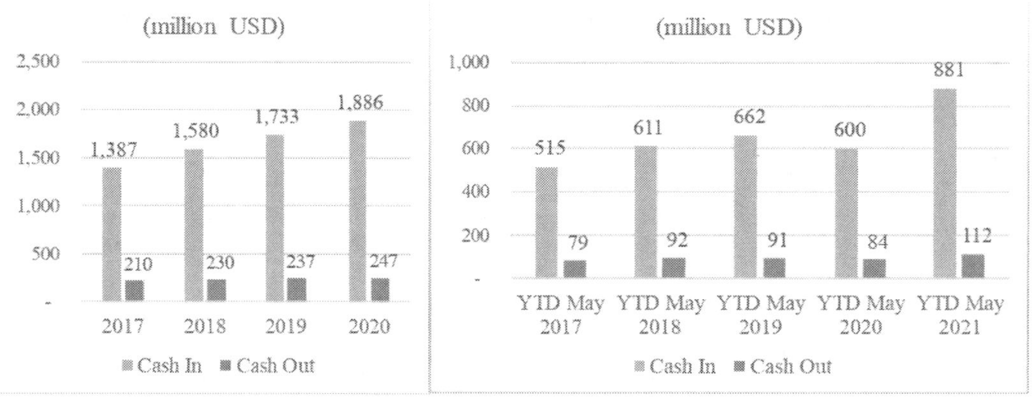

Fig 5. Money transfers the whole period of 2017-2020 and YTD May period of 2017-2021 [20].

According to the National Statistics Office of Georgia, for 2020, the turnover of SMEs in 2019 amounted to 44.5% of the total turnover. Noteworthy that 39.4% of the total turnover, which includes large, medium, and small businesses, comes to wholesale and retail trade, in which the share of SMEs is more than 50% [9].

According to the data, the share and structure of products manufactured in Georgia, 58.4% of the total value of products manufactured by enterprises in 2019 falls on SMEs.

According to the 2020 report of the National Statistics Office of Georgia, in 2019, the share of SMEs in added value was 59.3%, the share of employees was 65%, the turnover was 44.5%, and the share of output was 58.4% in total turnover. These numbers show how important SMEs are for the Georgian economy.

5 Conclusions

Beginning December 2019, events around the world related to the new coronavirus pandemic have clearly shown that the world was not ready for such challenges, and even developed and strong economies were unable to withstand the economic pressures caused by global disruptions in the supply chain. Small and medium-sized enterprises were especially vulnerable in this situation, resulting in production interruptions, temporary suspension, layoffs of employees, and change of the central business.

Although small and medium-sized enterprises are pretty "flexible" and can rapidly respond to external factors, they are remaining vulnerable and cannot overcome the crisis period independently and require active involvement and assistance from the government.

Since in the world and Georgia, a significant share of the economy falls on small and medium-sized enterprises and even more critical on the number of employees, delays in providing state aid to enterprises can cause significant losses and recession for the country's economy.

In addition to the fact that the leading countries of the world and their economies, as well as the economy of Georgia, were not ready for such a global challenge, which had a shocking effect all over the world, it should be noted that the private sector, especially in Georgia and especially in SMEs faced a new reality, after which they had to handle digitalisation of their services and offer remote services to their customers. Unfortunately, a crisis management system has not yet emerged in Georgia, and mostly the companies do not have an action plan for business continuous operating. In this regard, companies should identify risks raised and develop an action plan for their settlement. Foreign companies in developed countries have good experience in those mentioned above. Most of them have developed an action plan for business continuity and data recovery, which is also regulated by one of the International Organization for Standardisation standards, namely ISO: 22301. As per such a plan, companies simulate specific scenarios that identify significant risks and the economic and financial implications associated with them. Also, within the framework of this plan, it is possible to introduce a completely different activity to avoid complete cease of activity, such as a temporary "survival" plan, which the company will carry out in case the company cannot fulfil due to a decrease in demand.

In our opinion, besides the timely state financial aid for SMEs also is needed professional and technical support, in the symbiosis should be the companies' involvement to develop crisis management, risk assessment, and business continuity plan. In this regard, the state should provide SMEs with all the necessary information, which will raise the awareness of these enterprises and help them act rapidly and in agreement with the state in tandem, which will significantly simplify the process for both parties and the actions will be maximally coordinated. This will be relatively easy to achieve in Georgia, due to preconditions and essential elements for managing these processes already exist in the country, such as a precise classification of small and medium enterprises, as well as defining the main activities of enterprises with NACE codes and access to the Revenue Service web-portal for all registered taxpayers. By the mentioned web portal, it is possible to establish communication between the enterprise and the state. In addition to the above, since SMEs do not have enough financial, material-technical, and human resources to digitalise their services, it would be advisable if the state provides for SMEs with the internet platform, through which the companies could continue their activities and provide services to their customers via the Internet.

References

1. O. Prokopenko, C. Toktosunova, N. Sharsheeva, R. Zablotska, V. Mazurenko, L. Halaz, Problemy Ekorozwoju, **16(2)**, 7-17 (2021). http://doi.org/10.35784/pe.2021.2.01
2. United Nations, Micro-, Small and Medium-sized Enterprises (2020)
3. UN Global Compact, Helping Small Businesses Survive COVID-19 (2020)
4. I. Khlobystov, L. Horoshkova, V. Trysnyuk, D. Tarasenko, O. Tarasenko, L. Filipishyna, European Association of Geoscientists & Engineers, **1** (2021)
5. O. Prokopenko, R. Kornatowski, Marketing and Management of Innovation, **XIV(I),** 165–173 (2018). http://doi.org/10.21272/mmi.2018.1-22
6. M. Filipova, R. Yuleva-Chuchulayna, Revista Inclusiones, **7(3)**, 461-476 (2020). http://www.revistainclusiones.com/gallery/38%20VOL%207%20NUM%20ESPECIALJULIOSEPTIEMBREEE2020REVINCLUSIOO.pdf

7. EuroStat, Glossary: Enterprise size (2016)
8. EuroStat, Non-financial corporations - statistics on profits and investment (2020). https://ec.europa.eu/eurostat
9. National Statistics Office of Georgia, Euro SDMX Metadata Structure (ESMS). Enterprise Statistical Survey (Annual) (2019)
10. Organisation for Economic Co-operation and Development, OECD Policy Responses to Coronavirus (COVID-19) (2020)
11. International Monetary Fund, Special Series on COVID-19 (2020)
12. M. Chorna, L. Filipishyna, G. Nord, S. Tkachenko, K. Velychko, J. Secur. Sustain, **8**, 413 (2019)
13. O. Laktionova, V. Koval, N. Savina, B. Gechbaia, Bull. Georg. Natl. Acad. Sc. **15 (2),** 177-184 (2021)
14. PriceWaterhouseCoopers Georgia, Georgian companies facing the challenge of the Covid-19 pandemic (2020)
15. K. Nazarova, V. Mysiuk, V. Gordopolov, V. Koval, I. Danilevičienė, Business: Theory and Pract. **21**, 293 (2020)
16. World Trade Organization, Small business and trade (2020)
17. I. W. E. Arsawan, V. Koval, G. Duginets, O. Kalinin, I. Korostova, E3S Web Conf. **255**, 01012 (2021)
18. N. Dalevska, V. Khobta, A. Kwilinski, S. Kravchenko, J. Entrep. Sustain. Issu. **6**, 1839 (2019)
19. National Bank of Georgia, Financial Stability (2021). https://www.nbg.gov.ge

Smart Specialisation Strategy in the Post-Pandemic Times: the Evidence of Belarus

Yuliya Pranuza*

Francisk Skorina Gomel State University, Department of Economics and Management, 246028, Gomel, Belarus

Abstract. Innovation is recognised as one of the means of implementation for the entire 2030 Agenda for Sustainable Development. The Covid19 pandemic has resulted in slowed economic development. The innovation will have to play an essential role in recovering from the aftermath of the coronavirus. This article discusses how the application of the Smart Specialisation Strategy (S3) will allow faster recovery from the Covid-19 pandemic. The results of the study indicate that (1) S3 are place-based, innovation-led transformation policy for growth and sustainability; (2) the application of S3 allows to choose the priorities within the framework of innovative development; (3) Smart Specialisation can be an effective tool to help regions recover from the pandemic crisis and find new opportunities for more sustainable economies; (4) European Union (EU) countries have successful experience of the application S3; (5) the approach of Smart Specialisation can be applicable in Belarus for creating a competitive advantage for regions; (6) the areas of national and regional importance for the economy of Belarus were identified that could form the basis of the application of the Smart Specialisation approach in Belarus. Keywords: Smart Specialisation Strategy (S3), sustainability, Sustainable Development, Covid-19 pandemic.

1 Introduction

The World Health Organization (WHO) has declared a Public Health Emergency of International Concern (on March 11, 2020) due to coronavirus. The governments worldwide responded to the virus by severe locking down meant to flatten rates of infection.

COVID-19 has had severe negative social consequences worldwide. In addition to its impact on health, it has also significantly impaired economic activities in many countries. The pandemic has had a significant impact on businesses worldwide by lowering the demand for products.

In September 2015, the General Assembly of the United Nations adopted the 2030 Agenda for Sustainable Development to promote inclusive and sustainable economic development. The 2030 Agenda covers 17 Sustainable Development Goals (SDGs) and 169 related targets. Goal 9 calls to work together to 'build resilient infrastructure, promote

* Corresponding author: upronuzo@gmail.com

inclusive and sustainable industrialisation and foster innovation. Innovation policies are also implicit in goal 8, including target 8.2 'Achieve higher levels of economic productivity through diversification, technological upgrading and innovation, including through a focus on high-value added and labour-intensive sectors', and target 8.3 'Promote development-oriented policies that support productive activities, creation of decently-paid jobs, entrepreneurship, creativity and innovation…' [1].

Innovation can contribute to achieving these goals and targets, as it drives productivity growth and conserves scarce resources. Moreover, achieving most, if not all, of the sustainable development goals will require innovation.

The innovations are one of the most effective strategies that firms have at their disposal to combat the adverse effects of a crisis. The innovation provides opportunities to adapt and demand products/services following the theory of dynamic capabilities and management.

The Covid-19 pandemic causes dramatic health, social and economic consequences. At the same time, the Internet and other new technologies started playing a significant role in supporting remote working, e-teaching, online collaboration, etc. The pandemic and the related lockdown in many countries made digital work a new norm. It is establishing very effective networks for collaboration in the conditions of lockdown.

Innovation policy plays a pivotal role to tackle current dramatic social challenges. Today, the main challenge for all economies in the world is the Covid pandemic. Innovation and innovation activities should be the primary way out of the economic crisis.

To fully realise the potential of innovation for sustainable development, it is necessary to encourage and steer innovation efforts into areas critical for sustainable development and encourage the rapid and broad-based adoption and diffusion of innovations. Technology and innovation to achieve the development goals for territories allow to move and find new ways of solving current challenges of the pandemic.

The overcoming of the crisis triggered by the pandemic requires unprecedented cooperation and innovation capacities. Building efficient partnerships to deliver innovation processes is the essence of Smart Specialisation.

The present article tries to determine the list of sectoral manufacturing areas of Smart Specialisation for Belarus.

2 Smart specialisation

Smart Specialisation is a set of rules for choosing priorities within the framework of an innovative development strategy. These rules are introduced in a single EU guidance document [2].

Foray (2015) defined Smart Specialisation as 'the capacity of an economic system (a region for example) to generate new specialities through the discovery of new domains of opportunity and the local concentration and agglomeration of resources and competencies in these domains [3].

The rationalism of Smart Specialisation means that each region can find its own unique and justified way of development without copying the priorities of other territories. This requires focusing efforts and resources on a limited set of areas in which leadership can be achieved.

'Smart 'means 'unique 'concerning the strategy of the region. Uniqueness can be defined as follows: 'Being the only one of its kind; unlike anything else [4]. Uniqueness is one of the development priorities thanks to which the region's accumulated competencies and resources are channelled into new activities.

Smart Specialisation allows to identify a few areas with potentials for future growth and to define the associated transformation goals.

There are steps of developing S3 are: i) identification of priority areas (requires careful planning), ii) design of transformational roadmaps, iii) implementation with the help of an action plan. S3 priority areas consist of priority areas and transformation goals.

The main results of Smart Specialisation are structural changes that occur based on economic diversification associated with innovation. This expectation is based on growing evidence from many European and non-EU regions and countries where S3 has been used to support innovation for growth.

There is a widely shared expectation that S3 carry the substantial potential to effectuate regional and national ways to SDGs. The new concept for Sustainable Smart Specialisation has appeared [5]. It involves sustainability and inclusiveness and relates to the new EU priorities during the post-pandemic recovery. It is a research and innovation policy approach contributing to the achievement of 17 SDGs. The concept of Sustainable Smart Specialisation is a place-based agenda that aims at improving sustainability and inclusiveness through an innovation-driven policy.

3 International experience

Smart Specialisation plays an essential role in EU countries and regions. Currently, 187 regions of EU countries are registered on the online Platform of Smart Specialisation (Smart Specialisation Strategies Platform) [6]. These regions are developing and implementing strategies. In addition to the regional profile, the Platform identifies thematic areas of regional Specialisation to promote interregional networking among regions and to generate synergies.

The S3 Platform provides support in exploring and setting up international collaboration in Smart Specialisation. The S3 Platform identifies thematic areas of regional Specialisation to promote interregional networking among regions.

The European experience of Smart Specialisation has been recognised worldwide with a particular emphasis on local needs. In recent years, there has been a growing interest in adopting Smart Specialisation as a reference approach to a localised innovation transformation in some countries across the globe: 23 non-EU countries have been working on Smart Specialisation (e.g. Mexico, Australia, South Korea, Latin American, Montenegro, Serbia, Tunisia, etc.) [6].

In 2021 for the first time in Russia the Higher School of Economics conducted a comprehensive assessment of regional Specialisation based on the principles of smart Specialisation. 12 Ukrainian regions were registered on the S3 Platform [6]. The map of the modern economic Specialisation of Russian regions was compiled [7].

4 Methodology

The identification of Specialisation of regions is based on the method of determining significant cluster groups, what has been used to identify industries of Specialisation in the EU [8]. The development of the regions is determined by the specification of their trade with neighboring territories.

Industries are classified as 'traded' or 'local'. Traded industries are industries that are concentrated in a subset of geographic areas and sell products to other regions and nations. Local industries are present in most geographic areas, and primarily sell products locally.

Today there are different criteria for assigning industries to a particular category. This research is based on the approach of Michael Porter (Porter, 2003). As a criterion of trade, he used the uneven distribution of sectoral employment throughout the country [9-10].

The national level of the industry in the region is determined by the share of the region in the total sectoral employment in the country. Specialisation of national significance in a particular region is recognised as one of the sectors that make up the first 80 per cent of all sectoral employment in the country [7].

The industry of the regional level is determined on the basis of the ratio of the share of the industry in the region to the average national level. The industry of Specialisation of regional significance in the region is recognised as one in which employment is higher than the average employment rate in this sector in the country.

5 The evidence of Belarus

International experience confirms the importance of S3 in setting innovation policy priorities. This approach can also be used to determine the list of sectoral areas of Smart Specialisation in Belarus.

The main goal of the S3 is to help regions choose sound and non-intersecting development priorities of development in perspective fields of economic Specialisation. The list of such sectors in Belarus is generated [11], which can be the first step in this way.

The study is based on the approach of Michael Porter (2003), who used the uneven distribution of sectoral employment across the country as a factor for Specialisation [9-10].

The study identified sectors of national and regional level based on employment statistics (2018-2020) [12].

The industry of the regional level is essential at the national importance since the regions with relevant specialisation form a significant proportion of the country's total sectoral employment. This is characteristic, for example, for Manufacture of essential pharmaceutical products and pharmaceutical preparations in Minsk region (70% of country employment in the industry) and Manufacture of coke and refined petroleum products in Gomel region (28% of country employment in the industry).

The sectors of regional level are basic to the economy of a particular region. Their share in regional employment is higher than the national average level (with these regions contributing little to the country's sectoral employment). An example, Manufacture of essential pharmaceutical products and pharmaceutical preparations in Minsk (at 1.5% of national employment in this industry).

Among all the regions of Belarus, the economy of the Gomel and Minsk regions, the city of Minsk has become the most diversified (with a wide range of specialisations). At the same time, the choice of development directions of the region can be based on the number of industries of Specialisation and their type of level (national level, regional level, and national and regional level at the same time). These indicators help to assess the degree of diversification of the industry structure, the role of each industry in the economy of the region/country.

The growing number of industries of Specialisation depict the level of diversification of the region's economy and reflect the importance of each sector to the economy of the region and the country as a whole.

Diversified economies should search for niches in global value chains. For territories with few specialisations, interregional cooperation could be an optimal strategy, opening up new markets, opportunities for related activities and increased industrial diversity in the future.

6 Conclusion

Given the wide sectoral diversity, it is essential to focus investment resources and support measures on a limited list of identified priorities. The areas of Smart Specialisation in Belarus were determined. Industries of national and regional level (at the same time) for the economy of Belarus have been singled out: Manufacturing of food products and other non-metallic mineral products, Manufacture of machinery and equipment, Manufacture of textiles, Manufacture of rubber and plastic products.

A smart Specialisation is a tool for intensifying innovative development, structural modernisation and improving the competitiveness of regions. The Smart Specialisation policy can soften the Covid-19 economic crisis and promote a swift recovery.

Innovation will be critical for increasing the competitiveness of regional economic systems and developing new competencies of this foundation and recovery after the pandemic.

This work was supported by the Department of Economics and Management of the Francisk Skorina Gomel State University in Belarus (Gomel). I'd like to thank my scientific mentor Valeriy Sialitski, a well-known public figure in Gomel region and a former chairman of the Gomel Regional Council of Deputies for the opportunity for this scientific work to be completed.

References

1. United Nations. Transforming our world: the 2030 Agenda for Sustainable Development. A/RES/70/1 (2015)
2. Guide to Research and Innovation Strategies for Smart Specialisations. Brussels: European Commission. Available at: http://s3platform.jrc.ec.europa.eu/documents/20182/84453/RIS3+Guide.pdf/fceb8c58-73a9-4863-8107-752aef77e7b4/ (Accessed December 5 2020) (2012)
3. D. Foray, Smart Specialisation – Opportunities and Challenges for Regional Innovation Policy (Regional Studies Association, Routledge, 2015)
4. Unique. The Oxford English Dictionary, https://www.lexico.com/definition/unique, last accessed 2021/08/07 (2021)
5. N. Nakicenovic, C. Zimm, M. Matusiak, K. Stancova. Smart Specialisation, Sustainable Development Goals and environmental commons. Conceptual framework in the context of EU policy (Publications Office of the European Union, Luxembourg, doi:10.2760/766406, 2021)
6. Smart Specialisation Platform. Brussels: European Commission. Available at: http://s3platform.jrc.ec.europa.eu/ (Accessed October 10 2021)
7. V. Abashkin, L. Gokheberg, Y. Eferin et al.; L. Gokheberg, E. Kutsenko (eds.), Atlas of Economics Specialisation of Russian Regions (National Research University Higher school of Economics, Moscow, HSE, 2021)
8. Cluster Mapping Methodology. Available at: https://clustermapping.us/content/cluster-mapping-methodology/ (Accessed October 25 2021)
9. M. Porter, Regional Studies, 37(6-7), 549-578, DOI: 10.1080/0034340032000108688 (2003)
10. M. Porter, M. Delgado-Garcia, S. Stern, Journal of Economic Geography, 16(1), 1-38 (2015)
11. About the priority directions of scientific, technical and innovative activity for 2021-2025. Decree № 156 of May 7, 2020. Available at: https://president.gov.by/ru/documents/ukaz-no-156-ot-7-maya-2020-g/ (Accessed January 8 2021)
12. State Committee on Science and Technology. Available at: https://www.belstat.gov.by/ (Accessed October 22 2021)

Innovative Development of Public Policy on Food Security in the Period of the Covid-19 Pandemia

Larysa Hromozdova[1*], *Oleksandr* Yerankin[2], *Nadiia* Pysarenko[3], *Antonina* Artemenko[4], *Viktor* Maliarevskyi[1], *Yana* Dovhenko[5] and *Fato* Sharoian[6 7]

[1]Kyiv National Economic University named after Vadim Hetman, Department of Regions and Tourism, 03057, Victory Avenue, 54/1, Kyiv, Ukraine
[2]Kyiv National Economic University named after Vadim Hetman, Department of Marketing, 03057, Victory Avenue, 54/1, Kyiv, Ukraine
[3]Capital Union School, 03039, str. Keramichnyi 7-A, Kyiv, Ukraine
[4]Kyiv National Economic University named after Vadim Hetman, Department of Foreign of Marketing Faculty, 03057, Victory Avenue, 54/1, Kyiv, Ukraine
[5]Central Ukrainian State Pedagogical University named after Volodymyr Vynnychenko, Departments of Applied Mathematics, Statistics and Economics, 25000, str. Shevshenko 1, Kropyvnytsky, Ukraine
[6]Taras Shevchenko National University of Kyiv, 01033, str Volodymyrska 64/13, Kyiv, Ukraine
[7]National Bank of Italy, Via Nazionale str., 91, Rome, Italy

Abstract. The purpose of the article is to analyze the prospects for innovative development of public policy in the sphere of food security and to develop practical recommendations regarding the use of cultivated and wild plants in the sphere of food supply to improve the population's immunity during the pandemic. The article analyzes the impact of the COVID19 pandemic on the reformatting of public food security policy. The destruction of traditional food stereotypes of the population certainly took place. The emergence and development of innovations in the sphere of nutrition of Ukrainian citizens are considered. The tendency to increase the nutritional value of wild plants has been determined. Such plants are used to improve the immunity of the population and overcome the problem of lack of usefulness of food. The main barriers and problems of changing food stereotypes of the population and food policy of public authorities are substantiated. It is established that there is an insufficient level of support of public authorities in modern conditions of the development of the newest gathering in the sphere of food security in Ukraine. It has been revealed that the cost of manufacturing food products is relatively low in the modern world. However, the nutritional value and benefit to human health are minimal. There is a tendency to increase the number of diseases caused by the use of low-quality products. Alternative solutions to this urgent problem are considered. The influence of natural food additives and a return to the natural way of eating are investigated.

* Corresponding author: gromozdovag@gmail.com

The public authorities' importance of a broad information campaign regarding the dissemination of information to the public and food safety has been proven. As a result of the emergence of a pandemic, the public authorities must follow clear, consistent measures to improve the situation that has developed. It is definite that in addition to economic measures, attention should be paid to psychological measures aimed at changing the behaviour of the population. The necessity of destroying the industrial food traditions of the population was substantiated. The importance of actualization of the process of implementation of ecological food innovations was found out. Awareness of growing wild and gathering for food takes a prominent place. Ukraine has excellent opportunities for development in these areas. Reforming the state food security in today's conditions is necessary to preserve the health of the nation. The need to strengthen the role of public authorities in creating favourable conditions for environmentally sound use of food potential of wild-growing plants has been noted. Keywords: food, food policy of public authorities, wild-growing plants, reforming the food security system, food traditions and innovations, the newest gathering.

1 Introduction

The problem of food security in modern Ukraine has never lost its relevance. Recently, in conditions of relative prosperity, many citizens with excessive body weight have appeared. This is caused by overeating. It turned out that successes in combating manifest hunger provoked a new problem - "hidden hunger. Even a small number of calories in the food is accompanied by a lack of essential nutrients, microelements. Another danger was the contamination of food with a wide range of toxicants that enter due to general environmental pollution, chemicalization of agriculture, processing and food industries. It turned out that a decrease in its quality accompanies an increase in the quantity of food production.

However, the pandemic has threatened the quality and quantity of food supplies population. Risks of hidden and obvious hunger have increased. The economic crisis caused by quarantine measures contributes to reducing food production and the income of the population. Food is economically less accessible to consumers. This is especially true for high-quality, organic foods. High prices characterize such products. Therefore, the continuation of food traditions is unproductive. We need to rethink food traditions and find food innovations that will be relevant in the new reality.

2 Literature review

The pandemic phenomenon caused by the COVID - 19 coronavirus is being actively discussed in scientific circles. Such discussion is beneficial not only from the medical aspect but also from the economic, political, philosophical, social, psychological and other perspectives. In particular, the political perspective of solving social problems generated by the pandemic is devoted to the scientific reconnaissance of V.P. Bocharnikov, O.Y. Grynevich, I. Lazaridou, I.G. Markovich, S.V. Sveshnikova, A. Ostroushko, O.V. Polyakova, M. Stabulis and others.

Most studies cursorily note that the fight against the pandemic one of the negative consequences will be aggravation of the food problem in all countries of the world (K. Zubkova, I. Koval, O. Stoyanova). However, a unique analysis of the problems of aggravation of the food problem in the pandemic and the post-pandemic era is still lacking in the scientific literature. There is almost no discussion of the direction of the public policy of revising the food traditions of the past era. Not enough attention has been paid to the introduction of innovations in food production, trade and consumption.

Most experts note (E. Bragina, A. Gerasimenko, V.I. Kostra, T.V. Burlay) that there will be no return to the past in the post-consumer era. The life of mankind will change significantly. This means that changes will also take place in the sphere of nutrition.

Significant attention to the problem of reforming state food security in the context of traditional influences of the twentieth century is given in the works of T.G. Vasiltsiv, V.V. Boyko, O.I. Kutz, O.V. Bokiy, L.Є. Kupinets. One of the postulates of this approach was the focus on the use of agricultural products in nutrition. Products that were obtained from cultivated plants. Repeatedly expressed the idea of the high health and economic potential of gathering and food use of wild-growing flora (G. Bergson, A. Kashcheyev, A. Ursul). The potential of wild-growing flora remained almost unrealized.

At present, under the influence of the COVID - 19 pandemic, a rethinking of the world economy is rapidly taking place (V. Galasyuk, O.J. Grynevich, I.G. Markovich, M. Stabulis, I. Lazaridou, A. Ostroushko). Rethinking social consciousness, the formation of new realities (A. Gerasimenko, V.I. Kostra, T.V. Burlay). Consider innovative shifts in food stereotypes that need support from public policy bodies (K. Zubkova, O. Stoyanova, I. Koval) to ensure food security during the COVID pandemic, - 19. However, for the time being, the analysis of problems that hinder the use of the food potential of wild flora of Ukraine remains behind the attention. Such analysis is needed to solve the problems of food security during the pandemic.

The purpose of the article is to analyze the prospects for innovative development of public policy in the sphere of food security and to develop practical recommendations regarding the use of cultivated and wild plants in the sphere of food supply to improve the immunity of the population during the pandemic.

3 Results

Analyzing the problems of correcting food traditions and the formation of innovations at the public level, we will operate with the concept of "public policy" and "public administration". These terms are meaningfully broader in comparison with the recently popular terms "government policy" and " government administration" [14, p. 108]. This conceptualization takes into account the increasing role in the political process of non-state formations (local self-government, various institutions of civil society, non-state organizations, associations of citizens, private structures and agencies, etc.) that fulfill the role of partners.

The concept of "public policy" focuses on the fact that the policy of public administration bodies should not be directive, as it was thought before. Public policy should be based on the mechanisms of public coordination of the interests of stakeholders and aimed at achieving socially significant goals and social partnership in solving socially important tasks. Attention is drawn to the characteristic for the XXI century interaction of civil society and the state in solving those or other public problems. This position is a counterbalance to the diktat of state power, which was usual for the mentality of the twentieth and earlier historical stages of state creation.

Currently, the policy in the sphere of any, including food security, according to the new worldview is not imposed on society. The policy is implemented both through society and through state institutions of public authority. Trying to achieve the effect of the assemblage and readiness of society to the challenges, the state now uses modern methods of shaping public opinion. Under such conditions, public policy, regulated by democratic procedures, legitimizes public opinion. Public policy creates a counterpart in the person of the public, which is ready for dialogue and taking responsibility for compatible decisions in different spheres and, in particular, in the sphere of food security.

Scientists define the content of the concept of "food security" in different ways. Most often it is considered the ability of public authorities at the national level to guarantee self-sufficiency in basic products in amounts necessary to meet the food needs of the population and in a sufficiently diverse range, as well as the quality of food at a level that ensures the performance and health of citizens. No less difficult task of public administration than ensuring the physical availability of food is to create conditions for their economic accessibility. Otherwise, physical sufficiency, or even a surplus of food, will not protect more than a smaller part of the population from hunger [3, p. 132; 11, p. 265].

In addition to the physical and economic accessibility of food, it is advisable to highlight their cognitive accessibility. There are a large number of foods that are physically and economically accessible to citizens. But completely fall outside the food sphere through a system of cognitive gaps, cognitive barriers and food distortions of this individual, social group or society as a whole.

For example, exceptionally accessible in all attitudes are nettle, glague, dandelion, pigweed, purslane, amaranth and many other wild plants. However, for various reasons that are caused by socio-cultural factors, people do not perceive these very useful foods for food and even consider them forbidden [8, p. 7]. Therefore, the task of public authorities is not only to ensure the physical and economic availability of food to citizens. Public authorities must provide informative and cognitive information to the population about the usefulness of consumption.

Thus, a feature of public policy in the sphere of food supply of citizens is the expansion of nutritional competence of citizens. Dissemination of information on the use of the spectrum of cultivated plants and encouraging more people to consume currently little used plants (topinambur, turnips, borago). Supplement the food ration with wild plants (amaranth, pigweed, nettle, purslane, bluegrass, dandelion, elderberry, hawthorn, blackthorn, rowanberry ash).

The peculiarity of modern man's nutrition is that most of the dishes she consumes are not made as a result of home production, and is a product provided by the food industry in the form of finished products or various kinds of semi-finished products and the like [2, p. 50]. That is, about 50-55% of agricultural production is used in processed form, 20% is used in agriculture itself and only 25% is consumed fresh [6, p. 45]. At the same time, the expressive trend is that the fate of processed products is rapidly growing. Against this background, the issue of compliance with the quality and environmental safety of food is sharpened.

As L.E. Kupinets notes: "Identification of quality indicators of food and agricultural raw materials to the indicators of current standards usually indicates a decrease in the nutritional value of products and exceeds the maximum acceptable standards for the capacity of harmful substances in it, which does not give grounds to classify agricultural products as safe. After all, having achieved an increase in the quantity of food, most countries have lost considerably in its quality. Biochemical diversity of nutrition has been lost through a reduction in the biodiversity of crops and animal breeds. Modern studies in the field of food safety indicate that the processing receives raw materials of plant and animal origin with reduced indicators of usefulness" [5, p. 26]. Industrial processing of agricultural raw materials very often through intensive thermal, chemical and other influences reduces the impoverished presence of its useful substances.

In today's context, scientists conclude that almost 2 billion people in the world today suffer from "hidden hunger. Such a number is almost one-third of the world's population and has numerous negative health consequences. That is, the prevalence of "hidden" hunger and real hunger and malnutrition are of comparable proportions [12 p. 86]. One approach to eliminating nutrient deficiencies is to provide urgent assistance to vulnerable groups through pharmaceutical supplements. Globally, the consumption of vitamins, minerals and dietary supplements continues to grow. The U.S., EU, China, India, and Japan have the fastest growing markets for dietary supplements. Consumers in these countries are becoming increasingly interested in maintaining their health, increasing their ability to work, and decreasing the amount of time they spend on sick leave. Increasing drug costs are also contributing to this conclusion.

The second approach is to increase the micronutrient intake of the population through food fortification. After World War II, food fortification became common in the United States and some European countries. The advantages of food enrichment with simultaneous improvement of nutrition consist in the fact that traditionally consumed foods can be enriched without the need to change the behavior of food consumers, and the costs of enrichment are compared to lower [9, p. 112].

A third approach regarding overcoming "hidden" hunger through dietary diversity with a combination of bio-enrichment of staple food crops. This was made possible with the use of bio-enrichment through the development of new varieties of basic food crops with an increased content of nutrients. Bio-enrichment is a cheaper process than compensating nutrient deficiencies with food additives or enriching products during processing. Therefore, the introduction of bio-enriched products into the food system is very promising. Lack of dietary diversity is the main reason why most of the world's population does not suffer from micronutrient deficiencies in the diet.

In contrast to the problem of real, hidden hunger and malnutrition, there is the problem of overeating. This problem causes overweight and obesity. Obesity is a major cause of diabetes and cardiovascular disease and other illnesses. At the same time, much of the responsibility lies with food producers, who continue to produce products with high sugar and fat content, as well as abuse the addition of various flavor enhancers to food to increase sales and make super-profits [10, c. 43]. This situation is due to the fact that the interests of producers do not coincide with the interests of society. Therefore, in many Western countries with the support of the state began measures to reduce sugar, calories, fats in food products. To companies that do not comply with these measures, taxes and fines are applied. But in Ukraine, a large number of food safety indicators are not regulated and not controlled. At the same time consumption of low-quality food products is promoted not only by low solvency, but also by insufficiently high cultural level of the population.

In political terms, the issues of food safety, nutrition and food security are inextricably linked. Dangerous food causes disease and malnutrition, which particularly affects newborns and infants, the elderly and the sick. Disease of food origin is an obstacle to the economic development of the state. Because they put a strain on the health system and seriously harm the national economy, tourism, trade, etc. If we take into account the fact that nowadays food supplies are international, food safety has another global dimension [15, p. 22]. Therefore, it requires effective cooperation between national governments, producers, and consumers of food products to ensure food safety.

In addition, the issue of food safety has not only a global and national aspect, but also a personal one. Responsibility in the choice of products and the composition of one's own diet is the key to preserving health and preventing a whole range of diseases related to the way of eating. It depends on citizens [16, p. 122]. The population chooses food products, reckoning with the advertising of producers. The choice of products for consumption also formulates the request regarding the development of the food industry.

Lack of biologically valuable food, violations of its regime cause significant harm to human health, reduce physical and mental performance, increase the risk of disease, occupational injuries, nervous and emotional stress, increase the impact of harmful factors of production and radiation and reduce the period of active life.

It should be noted that as a result of coronavirus epidemic problems of food security and health of the nation become more acute. Due to the application of various quarantine measures, which limit the economic activity of agricultural enterprises, transport movement of food products, the work of some markets of rural products and public catering institutions give rise to risks in the quality and quantity of food consumption [13, c. 97].

During the deployment of the first wave of the pandemic over 5 million people were forced to go on leave after quarantine [4 p. 2]. Small and medium-sized business owners lost 25-50% of their revenues, causing 10-25% of employees to be laid off. Even large businesses lost 10-25% of profits and plan to downsize by 25% [7, p. 48]. Under the conditions of a sharp drop in income the population is forced to save even more and reduce spending, including on food, and switch from higher-quality to lower-quality food products, even more in violation of the balance of the food structure. For the most consumed food products in Ukraine (bread, sugar, oil, vegetables), which are characterized by inelastic demand. Meat and meat products, as well as milk and dairy products are characterized by high elasticity. The decrease in income of the population will lead to a decrease in consumption. This will lead to the transition of products with high elasticity to a cheaper group of consumption.

Taking into account the noted circumstances caused by the coronovirus, which will obviously become a permanent reality in the coming years, the President's advisor announced the creation of a special operative coordination council on food security under the working title "Harvest 2020". The council should work to stay ahead of possible threats to the food supply of the population. Because, through the decrease of immunity of the population can increase the loss of population from coronavirus disease. After all, declining food quality causes public health to deteriorate. First of all it concerns the immune system of the population, which already has an increased load as a result of other unfavorable factors, : environment, bad habits, stresses, infectious agents and the like. Therefore, with quantitative and qualitative nutritional deficiencies, a weakened immune system accounts for a high level of infectious and noninfectious morbidity.

Thus, in the context of a pandemic, the already acute problem of healthy nutrition is repeatedly exacerbated. However, practice expressively shows that it is not enough for public authorities to achieve success in ensuring the availability and economic accessibility of quality food. In addition to the presence of economic opportunities to purchase it is necessary psychological readiness of citizens to make a choice in the direction of a healthy diet. Also the prevalence of various eating disorders and destructive food stereotypes is growing among Ukrainians. In the sphere of food there is quite a wide range of untapped resources.It is wildly growing raw materials, the consumption of which is promoted by the ignorance of the population and the presence of a number of psychological barriers, which have deep cultural and historical roots. Overcoming this kind of psychological barriers to the food use of wild plants could help to solve the problem of food security in the fields of modern Ukraine.

In many scientific studies it was found that in wild-growing plants the spectrum and concentration of substances necessary for the human body is much higher than in cultivated plants, which were bred by breeding from wild-growing ancestors. It was found that the noted losses in the amount of nutrients and usefulness are the greater the further the process of cultivation of plants and animals has gone [17, p. 433].

The negative impact of industrialized agriculture on the nutritional quality of cultivated plants and animals has been established. Because many vegetables and fruits as a result of soil depletion over the past 30 years have reduced the content of useful substances by more than half. In addition, as a result of the use of agrochemicals, cultivated plants have acquired in their composition a high number of harmful substances, which were not in the preindustrial agrarian era in the products of organic farming.

It turned out that the agrarian domestication of nature and the emergence and development of the food industry creates food for humans in larger quantities than wildlife, but its quality becomes worse. The greater the amount of food, the worse its utility. That is, the transformation of wildlife into cultivated nature confirms the validity of Barry Kommoner's ecological law. The law notes that nature knows better, including in the area of creating adequate food for humans. Because of this scientific view of ecological and food problems, history and modernity have had the idea of reviving elements of the appropriation economy on a new basis. Such an approach was called the newest gathering [18 p. 4].

The newest gathering is understood not just as a revival of old gathering practices, but as a search for such forms of interaction between nature and modern society, where natural factors in their wild form are maximally involved in social activity. From this point of view it is the maximum use by the modern man of the gifts of wild nature in ecologically acceptable limits, the productivity of which is significantly underestimated in comparison with the production approach. This is a very promising direction of state stimulation of effective nature management, which needs preservation and expansion in the structure of the territory of the state of a part of the wild nature, part of which should be maintained absolutely reserved, and part of which should be set aside for the needs of the newest gathering.

Potentially the newest gathering can turn into a full-fledged sector of the economy. This implies a radical reform of the sphere of state administration towards the formation of the value status of wild nature in comparison to cultivated nature and the technosphere. The idea of the newest gathering can provoke the allotment of at least one third of the country's land area for wildlife. Also the ecologization of significant areas of the agrosphere, which will make it possible to harvest and use in nutrition a significant biomass of plants, which are currently neglected to qualify as weeds.

That is, in the conditions of modern Ukraine collecting can complement the agro-production. The newest gathering can be carried out territorially within a certain rural area (unused in the agricultural sphere of wild or at least semi-wild nature in the form of meadows, plantation forests, joists, gullies). Or grow wild plants in gardens and orchards. However, people call these plants weeds. Such weeds are actually valuable food and medicinal plants (pigweed, amaranth, purslane, quickweed). However, the revival of gathering to the level of an economic industry is not possible at present without the support of the state.

4 Conclusions and suggestions

The increase in the volume of food turned into a sharp impoverishment of the concentration of useful components in food and the emergence of a number of harmful substances. The new challenges of the post-industrial era have increased the search for artificial high-tech foodstuffs. And this is an increase in trends toward a return on a new basis to eating natural products, organic farming, and the use of forgotten wild plants in nutrition.

A particularly important task of public food policy in the new conditions is the removal of psychological barriers to the food use of wild plants. The usefulness of which has been proven by ancient folk experience and the latest numerous scientific studies. However, the refutation of traditional stereotypes of industrial food is very powerful. Therefore, the social stress created by the COVID pandemic has, in addition to the obvious risks, a positive consequence in the destruction of food traditions and the acceleration of the process of ecological food innovation. Among the innovations, the revival of public trust in wild plants and gathering is a significant one.

The prospects for further research is to develop ways of informational support of public authorities to ensure the availability of wild plants to the population of Ukraine and the normative-legal regulation of the rational latest collection.

References

1. Zakon Ukrainy «Pro osnovni pryntsypy ta vymohy do bezpechnosti ta yakosti kharchovykh produktiv» vid 23 hrudnia 1997 roku № 771/97-ВР. Vidomosti Verkhovnoi Rady Ukrainy (ВВР), 19, ст. 98 (1998)
2. H. Berson, Hydromeyoyzdat, 72 (1991)
3. T. Vasyltsiv, Stratehichna panorama, 2, 132–138 (2017)
4. V. Halasiuk, Ekonomika Ukrainy: pandemiia COVID-19, nova ekonomichna realnist, rekomendatsii. Ukrainian association the club of Rome. URL: http://www.clubofrome.org.ua/wp-content/uploads/2020/04/ DoslidzhennyaFond-Zajdelya.pdf.
5. A. Herasymenko, Visnyk Kyivskoho natsionalnoho torhovelno- ekonomichnoho universytetu, 3, 5–26 (2020)
6. O. Hrynevych, Ukrainskyi medychnyi chasopys, 3, 45–48 (2013)
7. K. Zubkova, O. Stoianova, I. Koval, Zbirnyk naukovykh prats ΛΌΗΟΣ, 48-49 (2020)
8. V. Kostrytsia, Ekonomika i prohnozuvannia, 1, 7–34 (2019)
9. A. Koshcheev, Pyshchevaia promышlennost. 258 (1981)
10. L. Kupinets. Ekonomika kharchovoi promyslovosti, 1, 43–50 (2009)
11. O. Kuts, Prodovolchi resursy, 14, 262–276 (2020)
12. L. Malakhova. Visnyk Kharkivskoho natsionalnoho universytetu imeni V.N. Karazina. Seriia: Mizhnarodni vidnosyny. Ekonomika. Krainoznavstvo. Turyzm. 1086(2), 8–89 (2013)
13. A. Ostroushko. Kreatyvnyi prostir. 1, 97–98 (2021)
14. O. Pukhkal. Investytsii: praktyka ta dosvid. № 24, 106–112 (2017)
15. S. Svieshnikov. Zbirnyk naukovykh prats Tsentru voienno-stratehichnykh doslidzhen Natsionalnoho universytetu oborony Ukrainy imeni Ivana Cherniakhovskoho. 2, 22–29 (2020)
16. Zh. Sokolova. Экonomyka, trud, upravlenye v selskom khoziaistve. 4, 122–132 (2019)
17. M. Stabulis. Zhurnal yevropeiskoi ekonomiky. T. 19, 3, 433–447 (2020)
18. A. Ursul . Khymyia y zhyzn. 11, 4–7 (1989)
19. Coronavirus: four out of five people`s jobs hit by pandemic. BBC news. 7, April (2020). URL: https://www.bbc.com/news/ business-52199888.
20. El-Erian M.A. Navigation the new normal in industrial countries (Washington, Per Jacobsson Foundation, 2010). URL: http:// www.perjacobsson.org/lectures/101010.pdf.

Problem-oriented Management of United Territorial Communities' Development Based on Decentralization under the Pandemic Influence

Yuriy Ivanov[1], *Olga* Ivanova[1*], *Viacheslav* Laptiev[2], *Olga* Polyakova[1], and *Viktoriia* Shlykova[1]

[1]Research Center for Industrial Development Problems National Academy of Sciences of Ukraine, Sector of Macroeconomic Analysis and Forecasting, Department of Macroeconomic Policy and Regional Development, 61145 Kharkiv, Ukraine
[2]Simon Kuznets Kharkiv National University of Economics, Department of Social Economics, 61145 Kharkiv, Ukraine

Abstract. The article is devoted to the problem-oriented management of spatial development of united territorial communities (UTC) under the conditions of decentralization under the influence of economic consequences of the Covid-19 pandemic. The article aims to justify the problem-oriented approach in the management of UTC based on the development of a cognitive map of the problems of their development and modelling the problem-oriented management of community resources on the example of human resources. The problem-oriented approach to resource management focuses on solving a problem as a specific theoretical or practical problem, which in turn adaptability, flexibility, the ability to make timely management decisions using available resources. The cognitive map of the four groups of problems of UTC development was developed: problems of the spatial development of UTC and administrative-territorial structure; problems of socio-economic development based on self-sufficiency; problems of local self-government system; problems of rural territories development. The example of human resources scenarios of problem-oriented management of community resources at the macro-, meso-; micro-levels were offered. The hypothesis that the effect of measures "from below" will spread more rapidly than "from above" has been proved.
Keywords: problem-oriented management, united territorial communities (UTC), problem-oriented approach, administrative-territorial structure.

The impact of the pandemic on the development of countries and their territorial units, entrepreneurship, and the development of human resources in general leads to two polar trends. On the one hand, these are the negative effects of Covid-19 on all spheres of life, associated with economic consequences for territorial communities, regions, businesses, unemployment for human resources. On the other hand, these are new opportunities for growth and development at all economic levels, which must be used to improve the competitiveness of all subjects.

* Corresponding author: laptevaou@gmail.com

Another problem in the development of united territorial communities is their formation as separate independent units, which are capable of self-development, but have many barriers: the distribution of authorities between the centre and united territorial communities, as well as responsibilities; problems of sufficient resource provision of self-development of territorial communities; insufficient level of skills of territorial communities management staff to develop effective strategies and development programs, real development projects, etc.

All this requires the introduction of an adaptive methodology in the united territorial communities development in a pandemic, which is based on a problem-oriented approach.

The article aims to justify the problem-oriented approach in the management of united territorial communities based on the development of a cognitive map of the problems of their development, as well as modelling the problem-oriented management of community resources on the example of human resources.

The economic crisis caused by Covid-19 changed the way of life and consumption of the population. Also, quarantine restrictions caused the transformation of economic players' behaviour in the production and consumption market on the principles of uncertainty and limitation. Both the pandemic and quarantine measures led to a slowdown in the economies of the world and Ukraine as well (Fig. 1) [1]. This negatively affects the economic security of countries and their sustainable economic development [2].

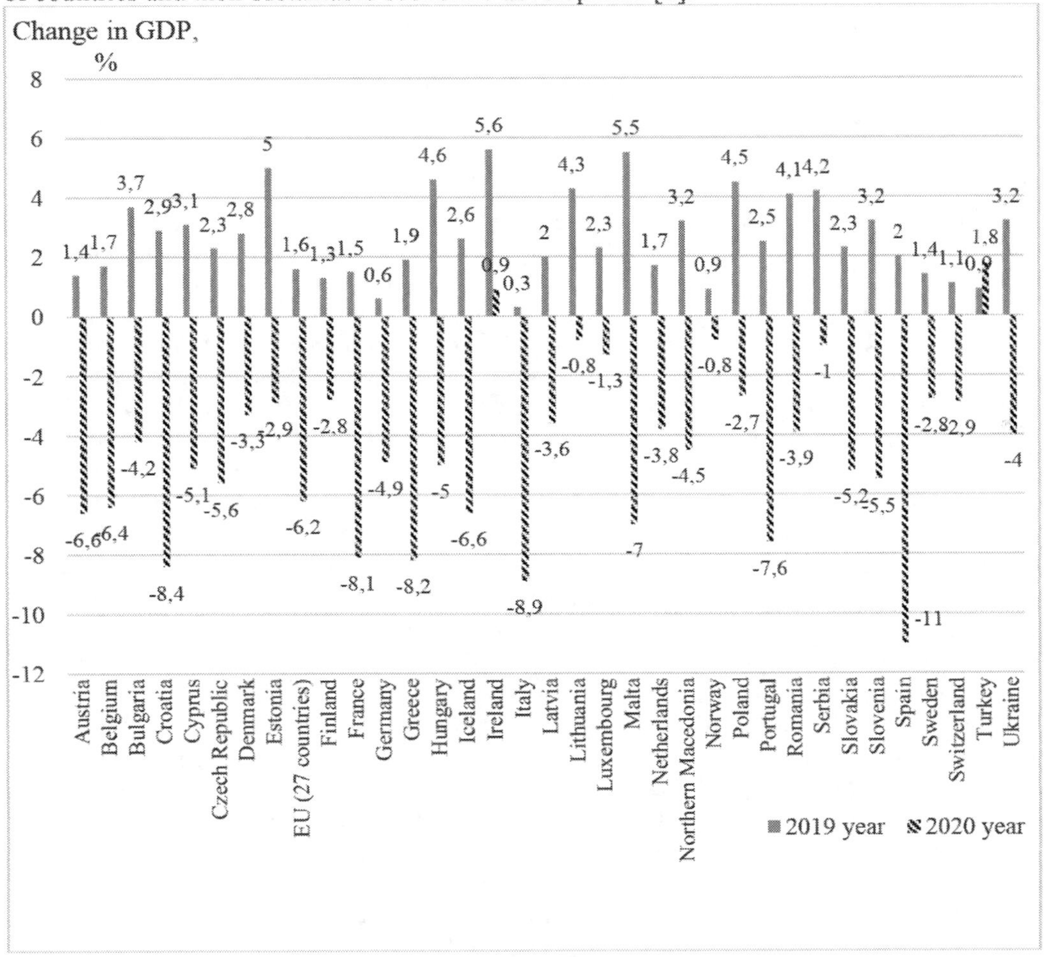

Fig. 1. Change in GDP in Ukraine and EU countries to the corresponding period of the previous year, %

The rate of GDP decrease is directly related to the negative consequences of the pandemic, so the most significant negative indicators are observed in Spain (11%), Italy (8.9%), Greece (8.2%), and Croatia (8.4%). In Ukraine, this indicator reached 4% due to the transformational nature of the economy, a large share of agriculture, poorly developed tourism, weak involvement in the production chains of added value in the global economy. Ukraine in 2020 showed fewer negative rates and better indicators of GDP dynamics than Hungary and Slovakia, which had a deeper recession than Ukraine at the end of the year.

The country's socio-economic development and regions depend on the possibility of territorial units of self-development based on resource self-sufficiency. Processes of reformation of economic and political processes in Ukraine occur under the influence of European integration [3]. It requires the development of reanimation measures to ensure sustainable social and economic development of regions based on the improvement of state and local resources management system in the sphere of tax [4] and budget relations between the centre and communities [5] human resources, etc.

Decentralization reform in Ukraine is a prerequisite for successfully implementing reforms in other spheres of territorial community development. However, its successful implementation requires the contribution of human resources.

The problem-oriented approach to resource management is "this approach focuses on solving a problem as a specific theoretical or practical problem, which in turn implies adaptability, flexibility, the ability to make timely management decisions using available resources [6]. Consequently, under the conditions of the decentralization reform, the application of this approach makes it possible to identify problems during the implementation of the reform and to involve effective methods and mechanisms of united territorial communities' development management of the executive authorities, public, business and population of the region to solve them.

Implementation of the problem-oriented approach requires identifying problems of united territorial communities' development based on decentralization under the pandemic influence.

According to the results of the survey by Swiss-Ukrainian Decentralization Support Project DESPRO [7] and the Ukrainian Association of united territorial communities [8] in 2020, 46 % of respondents noted the suspension of small and medium-sized businesses in united territorial communities (UTC) as a result of quarantine measures, 52 % noted the social responsibility of business in the fight against the pandemic, the focus on new consumer needs (21 %) and the shift to telecommuting (55 %). The situation of unemployment in the UTC by respondents is noted as critical because of the quarantine (57%), 44% of the residents of the UTC note the critical level of financial security, enough only for necessities [9].

Among the main challenges to local self-governance authorities and obstacles to development that befell Ukrainian territorial communities under the conditions of decentralization of power are the following groups of problems (Fig. 2):

1) problems of the spatial development of UTC and administrative-territorial structure: the lack of clearly substantiated criteria of unification of territorial communities; high fragmentation of administrative-territorial units of local level; unresolved issues of territorial planning and formation of general plans;

2) Problems of socio-economic development based on self-sufficiency: legislative failure to regulate the functioning of local Government on the distribution of powers, resources; low level of quality of life in the communities; low level of financial capacity of united territorial communities about delegated powers. Quarantine and suspension of work of many enterprises through COVID-19 caused problems in the economic sphere and affected the financial capacity of united territorial communities.

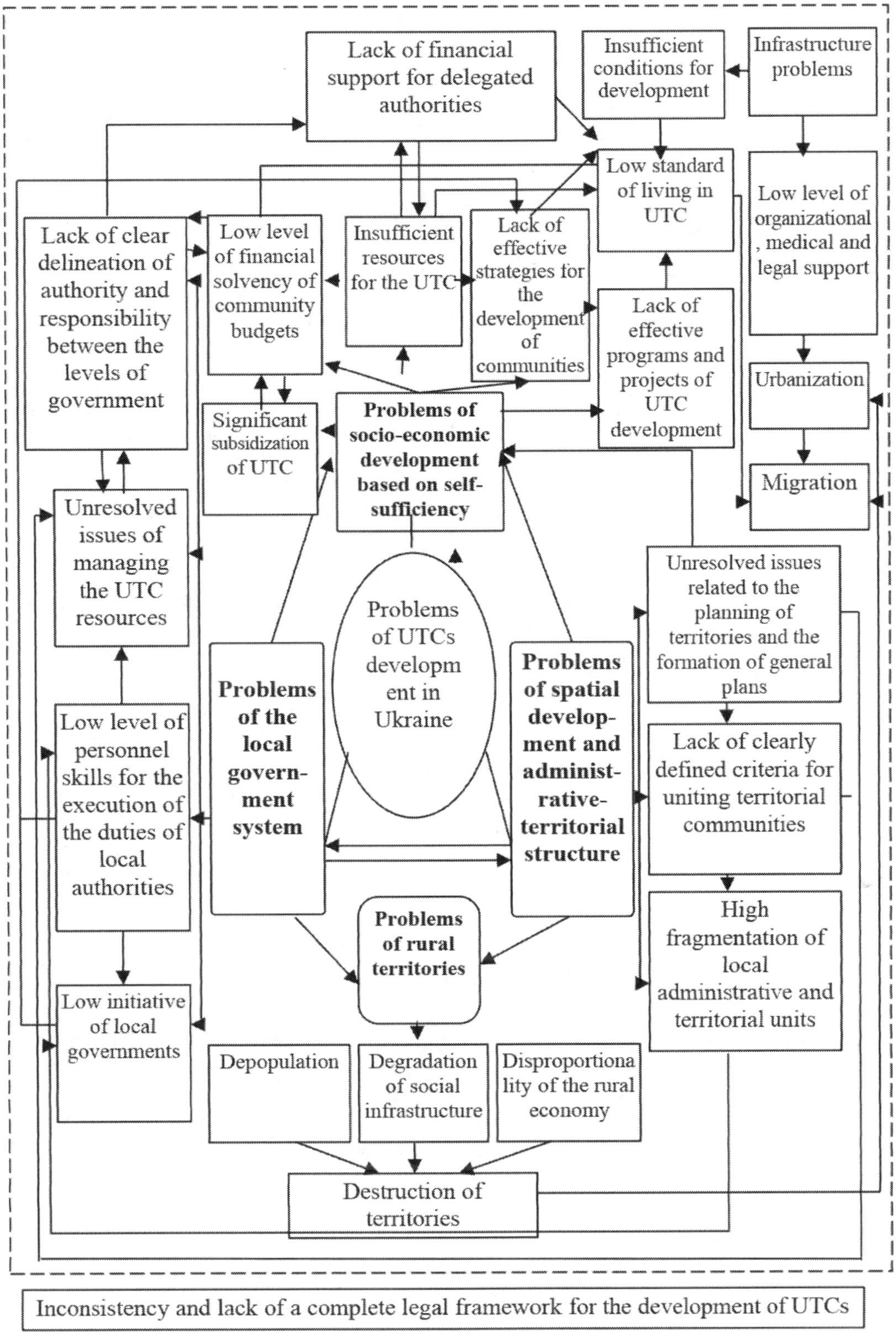

Fig. 2. Cognitive map of the relationship between the problems of the spatial development of UTCs

Because of the structure of own revenues of local budgets (approximately 95% of them are tax revenues, the benefits of which are offered to businesses in the regions) these budgets of territorial communities suffer significant losses. Thus, losses of regions due to anti-crisis measures proposed by the Government, according to the Open Budget portal and experts in financial and local budget management of DOBRE/USAID programs [10; 11], will amount to approximately 2-3% of tax revenues of general funds of local budgets in 2020;

3) problems of the local self-government system: low level of training, competence, awareness of personnel for the implementation of the powers of local self-government, human resources of territorial communities, lack of experience in negotiation, search for compromise and consensus, lack of systemic understanding of the process may hinder the cooperation of UTC for joint development, etc.;

4) problems of rural development economic problems caused by the specifics of rural UTC, inefficient methods of farming in rural areas; outdated material base, structural disproportions of production.

For effective implementation of reforms on decentralization, the human (in this case, staff) potential of local self-governments should be activated to form and implement appropriate mechanisms for coordination of territorial bodies of executive power. We are talking about self-governance as the ability of local governments to manage processes based on creating effective mechanisms for managing social relations in the region and ensuring effective interaction between subjects in the chain of relations "state-region-business-public" to ensure long-term socio-economic development of the region.

In order to model the scenarios, a simulation model of problem-oriented human resource development in Vensim was built (Fig. 3).

The indicators of the effectiveness of the proposed scenarios are the performance indicators: the value of the indicators of GDP per person, GRP per person, labour productivity; staff turnover; a number of persons who improved their qualifications

To identify the most effective options for the development of human resources at three economic levels, the following scenarios are considered:

- scenario 1 - "optimistic -1", which includes variables at the macro level: government spending on education and training (direct costs); expansion of the tax credit for individual entrepreneurs who are not the flat taxpayers, and expanding the list of services to include advanced training and retraining of both entrepreneurs themselves and their employees;
- scenario 2 - "optimistic-2", - changes in the costs of education and training programs for the development of human resources at the regional level and united territorial communities;
- scenario 3 - "optimistic-3" - management due to additional costs of enterprises for personnel [12].

Four scenarios were considered to identify the most effective options for human resources development at the three economic levels: the baseline scenario - unchanged indicators of problem-oriented human resource development at all economic levels (Table 1).

Each scenario has associated risks that need to be considered in the model. It is also essential to formulate criteria for the effectiveness of human resource development in terms of qualitative and quantitative parameters

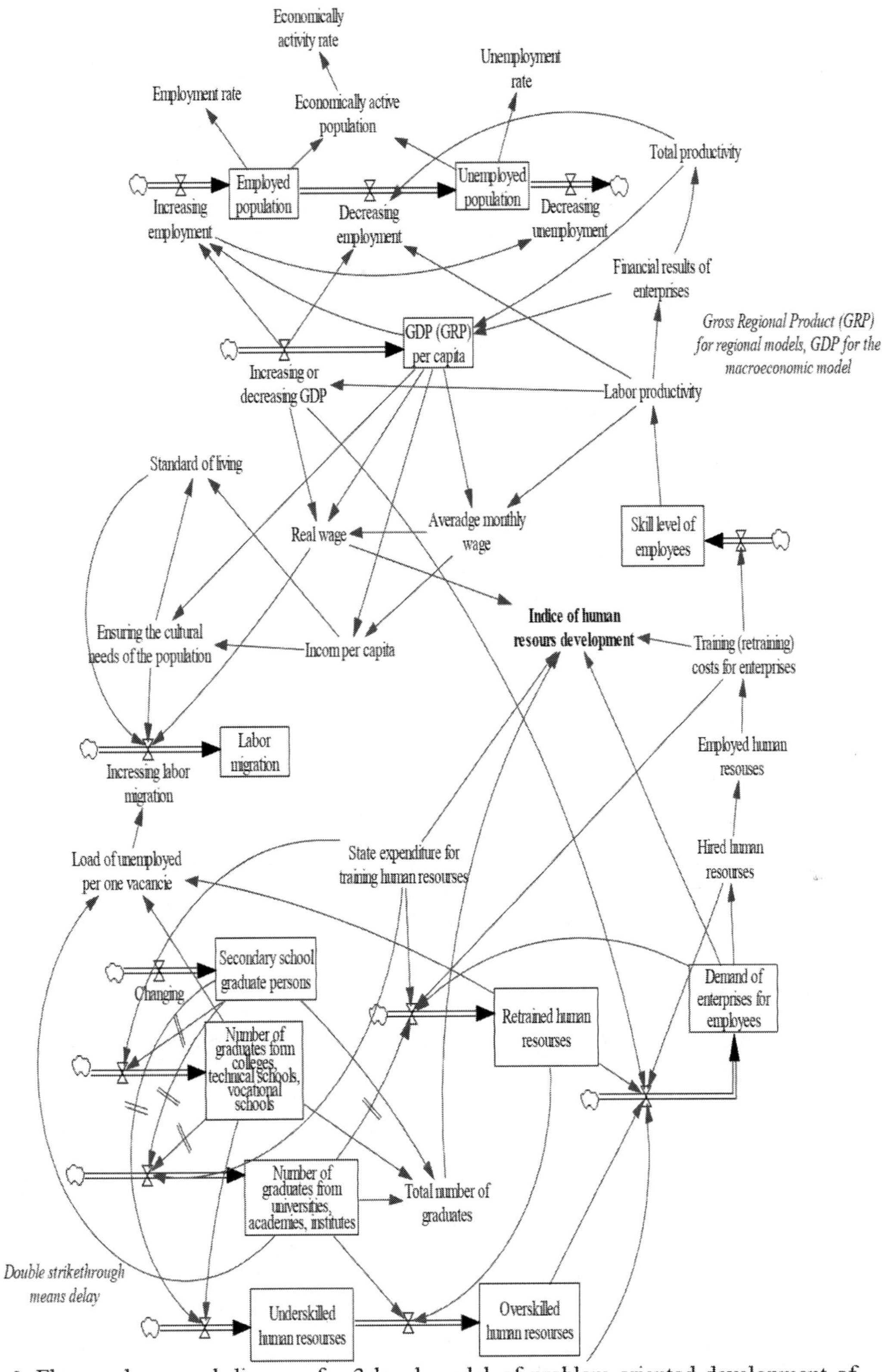

Fig. 3. Flow and a causal diagram for 3-level model of problem-oriented development of human resources

Table 1. Results of modelling scenarios of problem-oriented development of human resources

Scenarios	Scenario parameters	Performance for the national level	Performance at the mesolevel (region, united territorial community (UTC))
Scenario 1 (macroeconomic)	An increase of direct public costs for education to the maximum for the previous years (up to 6.78% of GDP); implementation of indirect costs by stimulating the professional development of private entrepreneurы and their employees by providing a tax rebate of 0.14 points	1.5 to 1.7% increase in GDP per person over five years compared to the baseline scenario	Increase of GRP per capita by 1.9% on average across the regions as compared with the baseline scenario (the most one in Chernivtsi Oblast (3.1-3.6%), the least one - in the Kyiv Oblast (0.9-1.0%)
Scenario 2 (mesoeconomic)	An increase in regional budgets' costs for education by 5%, with a continuation of the trend towards an increase in the share of regional budgets in the financing of education	An increase in total GDP by 2.3-2.8% over five years; an increase in total expenditures on education by 0.3 p.p. of GDP	Increase of GRP per person on average by 2.3-2.8% in the regions compared to the baseline scenario
Scenario 3 (microeconomic)	An increase in costs of the economic entities for personnel due to the costs of upgrading and retraining by 0.9-2.9% on average, compared with the baseline scenario	GDP growth per person by 0.56-1.1% compared to the baseline scenario	Increase of average monthly wages by 3.7% on average

Thus, the modelling of problem-oriented management of territorial communities based on decentralization was carried out using the example of human resources. The results showed that the improvement of particular indicators of problem-oriented human resource management under the micro-level scenario is achieved according to the forecasts faster than under the scenarios at the meso- and macro-level. The hypothesis that the effect of measures "from below" will spread more rapidly than "from above" has been proved.

References

1. Impact of COVID-19 on the country's economy and society: Results of 2020 and challenges and threats of post-pandemic development. Consensus forecast, 53 (2021) (in Ukrainian)
2. I. Gryshova, M. Kyzym, I. Hubarieva, V. Khaustova, A. Livinskyi, M. Koroshenko, Sustainability, **12(18)**, (2020)
3. V.V. Krivorotov, V.G. Mokhov, O.Yu. Ivanova, O.Yu. Polyakova, Bulletin SUSU MMCS, **13(3)**, 68-72 (2020)
4. A.O. Shapovalova, Yu.B. Ivanov, V.F. Tyschenko, V.V. Karpova, Journal of Tax Reform, **7(1)**, 68-86 (2021)
5. O. Ivanova, O. Poluakova, Journal of Tax Reform, **2(1)**, 25-42 (2016).
6. V. Laptiev, O. Ivanova, *Problem-based human resources management in the region. Coventry, United Kingdom*, in Perspective directions of scientific researches: Collection of scientific articles., 175–179 (2016).

7. Swiss-Ukrainian Decentralization Support Project DESPRO, URL: https://despro.org.ua/en/
8. All-Ukrainian Association of United Territorial Communities, URL: https://hromady.org/pro-asociaciyu/istoriya/
9. *Consequences of COVID-19 and quarantine measures for OTG. Survey results in Swiss-Ukrainian Decentralization Support Project DESPRO*, URL: https://decentralization.gov.ua/uploads/attachment/document/589/COVID-19_.pdf (in Ukrainian)
10. A. O. Dvihun, *Coronary crisis: new community opportunities and challenges for decentralization*, URL: https://niss.gov.ua/sites/default/files/2020-06/scientific-achievements-of-modern-society-niss.pdf (in Ukrainian)
11. State web portal of the budget for citizens, URL: https://openbudget.gov.ua/local-budget?id=26000000000
12. V. Laptiev, T. Stoyanova, S. Koev, P. Stoyanov, Z. Zhyvko, Academy of Strategic Management Journal, **18(3)**, 1-6 (2019).

Holistic Modelling Approach for the Management of Organisations

Harald Kitzmann[1,2]

[1]Bauman Moscow State Technical University, 105005, 2nd Baumanskaya St., 5, building 1, Moscow, Russian Federation
[2]Estonian Association for Renewable Energy, Estonia

Abstract. Developments in digitalization are in actual scientific discussion with leading focus on the automatizing of tasks especially in managers´ decision making, but also cover the view on the daily managerial task-fulfilment with digitalized solutions. Challenge for the managers is hereby the design of the digitalized work in a human-task-technology system by using advanced working models, methods and solutions. This compels to rethink causal-deterministic approaches and find new solutions, which especially occur in tasks of the strategic management. The purpose of this study was twofold. The first target was to identify and analyse the models and characteristics of strategic management decisions using approaches of the system theory, cybernetic management, and quantum theory. The second aim was to outline a framework of design parameters for creating causal-deterministic and holistic approaches in complex and complementary decision situations, so that they can be designed and processed digitally and implemented into human-task-technology systems, with outline the way how managerial decision-making will be changed. Keywords: complementary decision, dual process theory, strategic management, quantum decision theory, system theory, cybernetic management

It is proceeded as follows. First, the theoretical background to the study is outlined. It will be described the structural element in decision making as well as management of complex situations, paying special attention to complementary characteristics of decisions on the strategical and operational management level. Secondly, in the study it will be also examined different modelling approaches for solving complex and complementary decision situations. Building on this theoretical background, then, third it will be characterised the holistic, cybernetic modelling set-up and its transformation directions as appropriate approach for implementing into advanced digital solutions using the method engineering model. Then it will be discussed the derived framework as well as potential development areas in implementing the modelling approach, including the potential impact on decision situations and tasks, as well as impact on the needed competencies of users working with the improved solutions.

Considering the tripartite structure of decision making in a human-task-technology system, digitized solutions are not limited to the improving of the information processing, but also on the improving the dialogue interface, the improving of the competencies of users working with improved tools and the improved tasks and task solving in the organisations

[1, 2]. This transformation process and its impact on elements in the organisational knowledge system are shown in Figure 1.

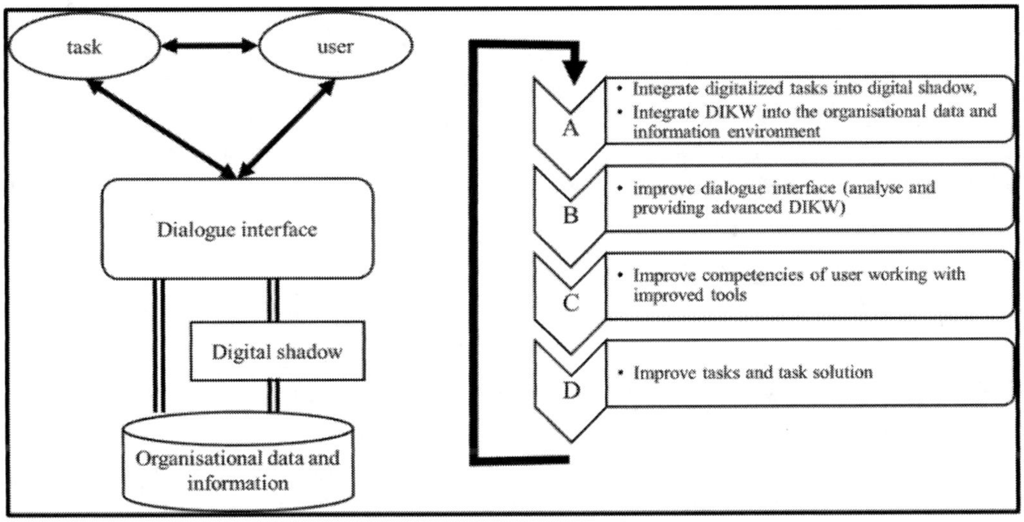

Fig. 1. Digitalization in the organisational knowledge system

Prerequisite for digitalization of management activity is the creation of theoretical concepts in supporting describing, explaining, prediction, simulation, decision, optimizing and gives orientation and reflection of the activities [3]. Hereby it should be considered modelling case, organization and situation-based individual, dynamical and process depending elements [4, 5], the modelling of the decision making-process on different management levels [6, 7, 8], the modelling of the cognitive process of intuition and deliberation in decision-making [9-15], as well as the tight integration of human decision activities and technology support [16-19]. These requirements find their implementation in advances holistic modelling solutions based on the dual-theory in decision making, the tri-partite modelling set-up, simultaneous management and dynamic screenplay approaches.

The EMO model (environment-management-organisation model) is a generic metamodel framework that allows to represent the participating, as well as observing and observed dimensions of cybernetic systems, but also allows to view, analyse and design situation- and decision-specific individualized elements as well as the whole system. In the model are described internal situations and processes on normative, strategic and operative level following different approaches of causality, correlation and complementary. The results in the management processes have impact on the element management, as well as on the interconnection with the environment and the design of the element organisation, which are designed as development modes, structuring forces, management practice, orientation level and value creation.

Understandings in strategic management consist of a time-dependent view with orientation to the future situation and a time-independent view focusing on the design of potential and the ways how to use the potential [3, 20-23]. The time-dependent view is discussed in cybernetics and systems theory with the present situation and their three attractors of development: optimizing equilibrium (stable equilibrium and steady state), instability and bounded instability. Schwaninger and Gälweiler developed a time-independent view in strategic management as a framework of control variables and identified depending on the management level differences in the purpose of these variables, their types and goals representing. The combination of time-dependent and time-independent view is already established in the Viable System Model (VSM Model) by Stafford Beer [20],

National Automated System for Computation and Information Processing (OGAS) [24], the Theory of Functional Systems [25], the sensitivity model by Vester [26], and the systemic economy [27]. But these models transfer the complex and qualitative characteristics into operational figures to handle them in deterministic-causal, data-driven analysis, without considering dualistic, complementary effects in decision making and processes in organisations. Therefore, the main understandings in strategic management were fragmented into different modules, whereas were found analogies and similarities in different scientific disciplines. Mainly these approaches are of a complementary character, but also extend traditional approaches with additional dimensions, and quantum mechanics approaches for different modelling set-up (Table 1).

Table 1. Approaches solving strategic management decision situations

Characteristics of complementary management	Attributes	Modelling approaches
Time-dependent view of development	Stability, optimizing, steady state, homeostasis	Causal-deterministic approaches, classical mathematics, physics
	bounded instability (alternating between stability and instability)	Alternating of causal-deterministic and non-causal dynamics
Time-independent view of development	Complementary of status and development direction of objects	Particle-wave approach in (quantum) physics
	Complementary of potential design and potential usage	Law on conservation of energy
Measures	Discrete and Extended measures on the operative management level	Flexibility cube; interval mathematics
	Non mathematical measures	"hermeneutic" interpretation
	Complex measure	One dimensional and multi-dimensional (e.g. Hilbert space)
Situation set-up and solving	Complementary of causal-deterministic and non-causal behaviour	Quantum decision theory
Integral set of management variables	Types interactions between three participants is reasoning the improvements	GHZ modelling set-up, Entanglement measure
	Integral set of objectives and actuating variables	Complementary cybernetics, System Dynamics
	inseparable unit of environment, process and scenarios	Screenplay approaches Dynamic Vector Logistics of Processes
	Integral set of elements and their connection	System theory

Beside the implementation of (weak and strong) Artificial Intelligence methods as subject in curricula, the interdisciplinary (especially from biological, chemical and quantum physic science) and cross-functional thinking and modelling, as well as the non-causal dynamic modelling, thinking in complex and fractal structures, dualistic thinking and modelling of combined deterministic and non-causal processes will be in the future found as learning outcomes in curricula and training programs. These aspects are challenging questions to be answered by national and international bodies (qualification frameworks, qualifications standards), professional bodies (occupational profiles and standards) and educational bodies (curricula and assessment specifications) as the stakeholders in the educational process.

The study shows that the complexity of decision situations allows using different proper solving methodologies with different reliable and suitable results, as well as enables and recommends continuous reviewing the design of decision situations. Furthermore, it underlines on one hand the organisational and time specific individualised character of strategic management decision situations. On the other hand, it revealed the reciprocal dependency between semantics, procedural model of the digitalized tasks and the user of the digitalized environment.

The research result closes the gap in finding implementation approaches for supporting the management in solving complex decision situations considering holistic system understandings and highlights the development areas for novel ideas and interdisciplinary transfer of knowledge developed in other scientific disciplines.

References

1. F. Fossen, A. Sorgner, Foresight and STI Governance, **13(2)**, 10-18. doi:https://doi.org/10.17323/2500-2597.2019.2.10.18 (2019)
2. S. G. Falko, V.V. Yatsenko, *Компетенции персонала в условиях цифровизации экономики. Информационное общество и цифровая экономика: глобальные трансформации,* 286-294, Krasnodar (2019).
3. J. Rüegg-StürmGrand, S. Das, St. Galler, *Management-Modell - Management in einer komplexen Welt* (2. Auflage ed.) (Bern, Haupt Verlag, 2020).
4. Jacobsen, L. *Bestimmungsfaktoren für Erfolg in Entrepreneurship* (Berlin, Freie Universität Berlin Hochschulschriften, 2003).
5. J. F. Urbánek, D. Král, *Modelling and implementation of added value controlling in small and middle enterprises continuity* (Prague, University of Finance and Administration Prague, 2014).
6. R. Stacey, *Managing the Unknowable: Strategic Boundaries Between Order and Chaos in Organizations* (John Wiley & Sons, 1992).
7. W. Ashby, *An introduction to cybernetics* (London, Champan & Hall Ltd., 1957).
8. H. Kitzmann, *Разработка модели управления гибкостью предприятия на оперативном и стратегическом уровне* (Moscow, 2018).
9. S. Milli, F. Lieder, T. Griffiths, *A Rational Reinterpretation of Dual-Process Theories* (2018).
10. G. Gronchi, F. Giovannelli, Front Psychol., **9**, 1237. doi:https://doi.org/10.3389/fpsyg.2018.01237 (2018).
11. V.I. Yukalov, D. Sornette, *How brains make decisions,* in Springer Proceedings in Physics. Universe of Scales: From Nanotechnology to Cosmology. **150**, 37-53 (2014).
12. M. Farjoun, Academy of Management Review, **35(2)**, 202–225 (2010).
13. A. Glöckner, C. Witteman, Thinking & Reasoning, **16(1)**, 1-25, (2010).
14. J. Evans, Thinking & Reasoning, **13(4)**, 321-339. (2007).
15. D. Marcial, M. Launer, International Journal of Scientific Engineering and Science, **3(12)**, 1-7, (2019).

16. C. Kruger, M. Snyman, *Analogy between Schrödinger's cat and Performance Measurement. Performance Measurement and Management: Public and Private* (pp. 393-400) (London, Performance Measurement Association, 2006).
17. H. Kitzmann, V. Yatsenko, M. Launer, Инновации в менеджменте, **1(27)**, 22-27, (2021).
18. M. Stöslein, P. Mertens, *Punktgenaue Unternehmenskommunikations mit Stakeholder Information systems.* DSWR - Datenverarbeitung -Steuern- Wirtschaft -Recht, 25-28. (2004).
19. L. Heinrich, A. Heinzle, R. Riedl, *Wirtschaftsinformatik* (Berlin, Heidelberg, Springer, 2011).
20. S. A. Beer, *Brain of the firm* (Wikey and Sons, 1981).
21. A. Gälweiler, *Strategische Unternehmensführung* (Campus Verlag, 1990).
22. M. Schwaninger, Systems Research and Behavioral Science, **18(2)**, 137-158 (2001).
23. R. Simons, *Levers of Control: How Managers Use Innovative Control Systems to Drive Strategy Renewal* (Harward Business School Press, 1995).
24. V. Glushkov, *Makroekonomicheskie modeli i printsipy postroeniya* OGAS (Moscow: Statistika Publ, 1975).
25. P. Anokhin, Uspekhi Fiziol. Nauk, **(1)**, 19-54 (1970).
26. F. Vester, *Die Kunst venetzt zu denken*, (dtv Verlag, 2015).
27. G. B. Kleiner, *Системная экономика: шаги развития* (Научная Библиотека, 2021).

Implementation of the Concept of Sustainable Development Based on Subsidiarity in Response to Current Dlobal Challenges

Lesya Korolchuk*

Lutsk National Technical University, Department of international economic relations, Lutsk, Ukraine

Abstract. With the current challenges of a globalized world, such as the crown virus pandemic and the growth of anti-globalization and anti-vaccination resistance movements around the world, the concept of sustainable development as a global strategy is under threat. There is a need to use such mechanisms and means of regulating global socio-economic processes that would ensure the effective implementation of sustainable development strategy while eliminating the acute perception of the planet's population of the negative effects of globalization. The article substantiates the application of the principle of subsidiarity in the implementation of the concept of sustainable development through the prism of an inclusive approach, as well as on the basis of the theory of sustainable development and deductive method of scientific knowledge it is developed a model for implementing the global concept of sustainable development on the basis of subsidiarity, which is to delegate the tasks of implementing the global concept from the global to the local level within the economic, social and environmental components of sustainable development and to specify these tasks at the level of the average citizen and local authorities. The implementation of the proposed model provides the involvement of every inhabitant of the planet in the processes of sustainable development with the ideological and financial support of local authorities in a decentralized environment, which will achieve sustainable development goals and reduce public resistance to global challenges.
Keywords: the concept of sustainable development, the principle of subsidiarity, an inclusive approach, anti-globalization movement, anti-vaccination.

The development of globalization in world economic processes, like anything new, has the experience of sharp criticism, rejection, and opposition. Thus, the hegemony of wealthy countries, the diligent work of transnational corporations and international organizations contributed to the emergence of the anti-globalization movement in the late 20th century. The comprehensive dictionary [1] interprets anti-globalism as a socio-political movement against international globalization, as in their opinion the negative consequences of this process far outweigh the positive ones.

*Corresponding author: l.korolchuck@lutsk-ntu.com.ua

Today, in the context of the crown virus pandemic, when the world is directly confronted with the challenges posed by globalization due to the increasing openness of national economies and international population mobility, the anti-globalization movement is gaining momentum. Many human casualties and the apparent helplessness of the world community in countering the spread of the virus have contributed to growing public distrust of the government and international organizations in particular.

Against the background of anti-globalization, a new branch is emerging anti-vaccination [2], as a broad civic movement that opposes vaccination against the coronavirus and denies its effectiveness and legitimacy. Based on the prevailing theory of global conspiracy, this state of affairs distorts the whole concept of sustainable development as the primary condition for the successful implementation of which see the reduction of the world's population. Anti-vaccines see the injection against the coronavirus as a means of combating the world's elite overpopulation.

Despite the above, the global concept of sustainable development must be best implemented to save our planet for future generations. We must be able to effectively control the crown virus pandemic globally to eliminate danger today and gain the ability to counter common threats in the future.

Thus, in the current conditions, it is necessary to apply such mechanisms, approaches and tools that would ensure the effective implementation of the concept of sustainable development on the one hand, as well as eliminate the sharp perception of the planet's adverse effects of globalization on the other. In this case, it is essential to transfer the relevant tools from the global plane, often immeasurable and far from the individual, to a lower local level - accessible, close and relevant to human life.

In the context of the above, we propose to apply the deductive method and inclusive approach, based on the theory of sustainable development, according to which the concept is based on three components [3]: economic, social and environmental.

Thus, as a method of scientific knowledge, the deduction will help us move away from the general global macroparameters of sustainable development, namely from its economic, social and environmental components to their micro manifestations at the local level.

In turn, the term "inclusion" is borrowed from the social sphere and confidently integrated into the economy at the beginning of the 21st century, and can be interpreted according to E. Goldberg [4] and E. Pacetti-Garr [5] as ensuring equal participation of people in socio-economic activities, both at the state and global levels, in the implementation of the global concept of sustainable development based on justice and stability, in order to improve the quality of human life.

Thus, to ensure an inclusive approach to implementing the global concept of sustainable development, it is necessary to delegate various types of socio-economic activities from the global level to the local within the components of sustainable development. Such delegation provides the principle of subsidiarity, which F. Barca [6] interprets as the distribution of tasks according to activities between management levels, depending on where these tasks will be performed most effectively.

It follows that the global concept of sustainable development based on subsidiarity should provide for the transfer of tasks to its local levels within the economic, social and environmental components of sustainable development to mitigate the negative effects of globalization and to ensure a decent standard of living with our planet for future generations.

It should be noted that the implementation of the global strategy of sustainable development based on subsidiarity in favour of the local level justifies itself through more significant opportunities for local authorities to apply an inclusive approach to regulating socio-economic processes in the global world. As a result, every citizen will feel significant and involved in the world community and global ideology, will be able to feel the results of the global strategy in their lives, fear and mistrust will disappear, resistance to globalization will decrease, and humanity will be able to meet modern global challenges effectively.

The above allows us to model the implementation of the global concept of sustainable development based on subsidiarity, the results of which are shown in Figure 1. Thus, Figure 1 shows that sustainable development goals at the local level are specified in economic, social and environmental components based on a deductive approach to the theory of sustainable development.

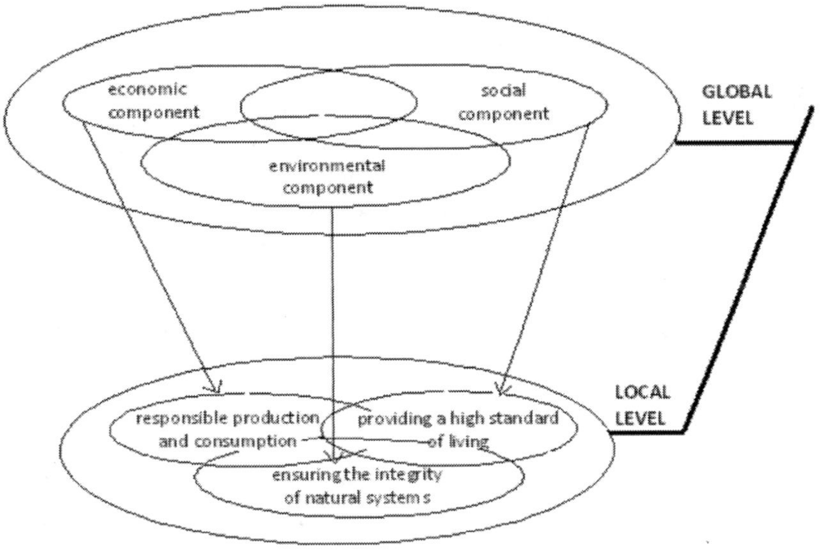

Fig. 1. The model of realization of the world concept of sustainable development based on subsidiarity

If the economic component of sustainable development at the global level involves optimizing the use of natural resources in production processes and maintaining economic growth of the economic system in terms of resource scarcity, then at the local level for the average citizen according to an inclusive approach is the task of responsible production, as a manufacturer-entrepreneur or production manager, by implementing the latest technologies, know-how and eco-innovations into production processes, saving resources and energy, eco-focusing product marking, extending the life cycle of goods, etc., as well as responsible consumption as a consumer of goods and services, by differentiating them according to the environmental criterion, cultivating a culture of economical and re-consumption of goods, sorting garbage and intelligent behaviour with waste, saving energy in everyday life, etc.

In turn, the social component of sustainable development at the global level involves human centrism and maintaining the stability of social and cultural systems, for which at the local level, the average citizen needs to take care of their health primarily in a pandemic crown virus, to be actively involved in sports and health sphere, to live and work in a stable and fair socio-economic environment with respect for fundamental human rights and freedoms, to have unhindered access to quality education and medicine, to grow and be educated based on freedom of choice, empathy and respect for people and nature.

The ecological component of sustainable development, which at the global level involves the preservation of our planet to ensure our livelihoods and for future generations of humanity, at the local level, the average resident faces the task of preserving the integrity of natural systems and species by caring for nature, flora and fauna in particular, introduction of ecological culture, active public work in the field of environmental protection, care for cleanliness and comfort in the native land, taking into account the environmental factor in all types of human activities, etc.

To fulfil the above tasks of implementing the concept of sustainable development, the average citizen at the local level needs comprehensive support from local authorities, as ideological through the formation and implementation of appropriate strategies for socio-economic development of regions, publicity and promotion of healthy lifestyles, ecological life, ecological consumption, etc. and material through the financing of relevant projects and programs in various fields, economic incentives for sustainable development, development of quality infrastructure to ensure a decent standard of living, support for disadvantaged and low-income citizens, etc.

In conclusion, to increase the effectiveness of the global strategy for sustainable development based on subsidiarity in favour of the local level requires decentralization of power in nation-states, which should result in the delegation of power and financial freedom to regional and local levels. In Ukraine, such reform has taken place and is now successfully nearing completion. Therefore, under favourable conditions, it is time to develop relevant tools for an inclusive approach to regulating socio-economic processes globally, such as sustainable development.

References

1. J. Shveda, Political parties: Encyclopedic dictionary, 500 (2004). URL: https://web.archive.org/web/20131109163554/http://www.lnu.edu.ua/faculty/Phil/Shweda_slovnyk.htm
2. M. Ifan, *Covid anti-vaxxers: "Shut down fake news sites," begs daughter*, BBC News, 6 August (2021). URL: from https://www.bbc.com/news/uk-wales-58103604
3. M. Kahn, *Sustainable development: The key concepts, issues and implications*, in Proceedings of the *I*nternational Sustainable Development Research Conference, 27–29 March 1995, Manchester, UK, Sustainable Development, **3**, 63-69 (1995). URL: https://onlinelibrary.wiley.com/doi/pdf/10.1002/sd.3460030203
4. E. Goldberg, Amer. For. Pol. Inter., **6**, 344-345 (2012).
5. E. Pacetti-Garr, *The Five Characteristics of an Inclusive Economy: Getting Beyond the Equity Growth Dichotomy* (Rockfeller Foundation, 2016). URL: https://www.rockefellerfoundation.org/blog/fivecharacteristics-inclusive-economy-getting-beyond-equity-growth-dichotomy/
6. F. Barca, *An agenda for a reformed cohesion policy. A place-based approach to meeting European Union challenges and expectations*. Independent Report prepared at the request of Danuta Hübner, Commissioner for Regional Policy, 219 (2009). URL: http://www.ecostat.unical.it/Dorio/Corsi/Corsi%202017/Politiche%20Sviluppo%20Locale/Materiale%20poleco/report_barca_v0306.pdf

Insolvency Law Adaptation to the Covid-19 New Challenges: the Process of Destigmatisation and the Focus on the Rehabilitation of Business

Elena S. Koshcheeva[1*], *Kamilla* A. Minnullina[1], and *Julia* N. Folgerova[1]

[1]Vyatka State University, Kirov, Russia

Abstract. The institution of insolvency (bankruptcy) occupies a special place in the state institutions system to support business entities in the recovery process from the crisis, given the selection mechanisms and rehabilitation opportunities inherent in it. Bankruptcy on the global economy scale serves as one of the anti-crisis regulation tools to improve financial and economic stability. The Covid-19 pandemic has actualised the ensuring the bankruptcy law institutional capacity problem by maximising its anti-crisis potential. Given its spread scale, the insolvency law adaptation to the Covid-19 new challenges is a challenge faced by all legal systems without exception. The findings of this study reveal that fragmented and uncoordinated national responses have common trends caused by the unity goal of minimising economically and financially destructive effects of the pandemic. This study objective was to identify the most effective strategy for the bankruptcy institute modernisation, which solution is based on the elements comparative analysis and convergence methods. In conclusion, the bankruptcy institute development basic tendencies and patterns have been revealed. Their analysis has allowed designating the legislation perfection most effective and perspective strategies in this sphere, which are determined by the course to aim at its anti-crisis potential development.
Keywords: Covid-19 new challenges, the process of destigmatisation, rehabilitation, insolvency, bankruptcy.

1 Introduction

The covid-19 pandemic had a negative impact and revealed vulnerable points in the social and public health spheres; it is also paralysed business activity and had unprecedented economic consequences, comparable in scale to the crisis of the 1930s [1]. The phenomenon of global economic interdependence determines the tendency towards similar responses to the spread of the virus eventually. The phenomenon referred to as the «copycat coronavirus policies» resulted from regulatory emulation [2]. This study demonstrates that these processes are also reflected in the institute of bankruptcy.

The article attempts to systematise the responses implemented by states to minimise the pandemic negative economic effect, directly reflected in the insolvency institute legal

*Corresponding author: usr22709@vyatsu.ru

regulation, to identify patterns and trends. This study objective was to identify the most effective strategy for the bankruptcy institute modernisation, which is based on the comparative analysis and convergence methods.

2 The institute of bankruptcy as a tool of anti-crisis regulation

In conditions of economic downturn, the legal regulation of insolvency deserves special attention from the point of view of its anti-crisis potential. The most dynamic development of legislation in this area is during the crisis. Given its spread scale, the insolvency law adaptation to the Covid-19 new challenges is a challenge faced by all legal systems without exception. Insolvency laws have changed in Germany, Great Britain, Italy, the USA, Poland, Colombia and other states, including Russia [3]. This article attempts to systematise and structure the Government's response at the national level to the institution of insolvency, aimed at minimising the negative consequences of the pandemic. The analysis allows us to demonstrate that disparate, uncoordinated measures have common trends and patterns, which are determined by the course of anti-crisis potential development of the legislation in the field of bankruptcy (insolvency).

The institution of insolvency (bankruptcy) occupies a special place in the state institutions system to support business entities in the recovery process from the crisis, given the selection mechanisms and rehabilitation opportunities inherent in it. Bankruptcy on the global economy scale serves as one of the anti-crisis regulation tools to improve financial and economic stability [4]. This special institution importance is determined by its main functions: bankruptcy prevention and avoidance under temporary economic downturn conditions; corporate solvency restoration with economic feasibility and viability to viable entities, through the crisis exit tools provision; the insolvent entity liquidation takes into account all parties (stakeholders) interests, which implies the debtor's assets effective and fair distribution.

On this basis, the bankruptcy institution provided the mechanisms inherent in it are adequate, can act as a powerful tool to protect business entities affected by the COVID-19 pandemic negative economic consequences. It is worth agreeing with the opinion that «bankruptcy is not a problem, but a solution especially for large companies» [5]. In addition, pandemic crises pose the massive corporate bankruptcies threat and catalyse inefficient insolvency mechanisms problems. Based on the above, the Covid-19 pandemic has actualised the ensuring the bankruptcy law institutional capacity problem by maximising its anti-crisis potential.

3 The bankruptcy institution transformation to minimise the pandemic negative economic impact

3.1 Emergency measures to prevent bankruptcies a massive wave

The pandemic response first phase was characterised by fragmented, uncoordinated measures designed to prevent a massive wave of bankruptcies and to allow time for businesses to recover [6]. Measures at this stage were time-limited and subject to cancellation (or review) depending on the pandemic scale and impact in a particular country.

One of the first measures taken in jurisdictions a number is the provisions' suspension imposing an obligation on controlling persons to initiate bankruptcy proceedings if there are objective insolvency indications. Most states suspended bankruptcy procedures initiation by creditors, including Russia, the so-called moratorium on bankruptcy. However, concerning creditors' rights to apply for bankruptcy, states a number followed the not prohibition path but correction towards insolvency threshold increase concerning the claims' amount and periods during which they are not fulfilled. In Singapore, for example, the amount of insolvency was increased tenfold and the period within which the debtor failed to meet the claim was doubled. Indian law initially restricted the creditors' rights to initiate bankruptcy proceedings, but a moratorium was subsequently imposed [7]. Despite the different approaches, in essence, they are defined by a common objective, which is characterised as taking into account the negative pandemic impact in assessing the introducing bankruptcy procedures reasonableness concerning a person.

The first phase emergency response measures, although fragmented and uncoordinated, aim to prevent bankruptcies from a massive wave by giving business entities a certain period to recover. There is no denying the differences in the ongoing reforms some elements, for example, most countries took the existing regulations optimisation and modification path (on a temporary or permanent basis), in other states, support measures were taken by introducing special regulations. Although individual elements differed from jurisdiction to jurisdiction, in essence, they all had the same objectives, by restricting or prohibiting the bankruptcy procedures initiation.

The moratorium is «postponement a kind» concerning entities with insolvency signs due to the circumstances exceptionality and essentially a rehabilitation mechanism. However, its introduction has aggravated the debtor and creditors balancing the interests' problem. In particular, it has brought to the fore such a measure admissibility issues in dishonest debtors respect, as well as its rationality in person respect, whose crisis financial condition arose before the pandemic. In Russia, to the difficulties associated with identifying which companies have been affected by the pandemic, the Supreme Court explained that countering to abuse these benefits by the debtor is possible only on the mechanisms «ex-post» basis, directly in the bankruptcy case, after the moratorium lifting.

The measures in this phase were characterised by a time limit. Due to the termination of their operation in most states, their consideration does not seem relevant. Moreover, their effectiveness from the standpoint of economic stabilisation is already being evaluated by experts [8]. At the same time, already at this stage, the processes of convergence of legal regulation are manifested due to the modelling and implementation of the most effective strategies for responding to the spread of the pandemic. The so-called response emulation, which consists in the similarity of response measures, despite its preferential consideration in the context of the social sphere, is thus reflected, among other things, in the legal regulation of insolvency.

Following this approach, the further development of legislation in this area is also characterised by common patterns that allow us to identify common trends.

3.2 Subsequent reforms to «smooth the insolvency curve.»

The measures taken at the initial stages had a stabilising effect. However, the crisis phenomena determine the reforming of the legislation on bankruptcy necessity in increasing its anti-crisis potential direction. Experts predict an increase in the corporate' bankruptcies number, which determines the legislation ensuring practical legal regulation task in this area, which provides for procedures optimisation and its unique mechanisms adaptation to new challenges.

The response second stage is already characterised by greater focus and consistency, which main task is to smooth the bankruptcy curve. This stage implies reforming the legislation on bankruptcy to increase its anti-crisis potential, optimisation, adaptation to the crisis phenomena caused by the pandemic.

As the Covid-19 spreading a result, the real economy sector and, in particular, small business entities turned out to be in the most vulnerable position. This specificity determined the general tendency to strengthen the pro-debitor component and the focus on increasing the efficiency and effectiveness of rehabilitation mechanisms.

3.2.1 Improvement rehabilitation procedures effectiveness

Given that the bankruptcy anti-crisis potential is primarily determined by the rehabilitation procedures effectiveness, their improvement is one of the law reform main trends, through procedures simplification and conditions for their introduction. Thus, in countries, a number (the USA, Singapore) the rules on thresholds of creditors' consent for the procedures introduction and financial reorganisation plan approval were revised or abolished. Improving the effectiveness of rehabilitation procedures is based on providing debtors with some of the advantages (especially in terms of flexibility and minimal court involvement) [9].

Low-efficiency the rehabilitation procedures problem in the Russian legal system, which is caused by pro-creditor concept prevalence. Reforming the legislation on insolvency in strengthening its rehabilitative orientation direction was and remains a topical issue, widely discussed in the doctrine [10]. It is necessary to admit that the crisis phenomena caused by the pandemic accelerated its resolution. At the moment the large-scale reforms draft to the legislation on bankruptcy is under consideration, providing radical changes of rehabilitation procedures.

3.2.2 The preventive restructuring mechanisms introduction

Due to the high costs of bankruptcy cases, the development of bankruptcy prevention tools is one of the urgent tasks. In the mass bankruptcies forecasts view, one of the reform key areas is characterised by the preventive restructuring mechanisms introduction. These tools should help to reduce the burden on the courts, as well as reduce the costs associated with bankruptcy proceedings. Recommendations to introduce preventive restructuring mechanisms in the insolvency institution were developed by the European Union back in 2019, but the crisis phenomena prompted lawmakers to enshrine them legally [11]. European states a number are adopting special rules on restructuring, providing for the possibility for entities in financial distress to restructure debts through an agreement with creditors outside the bankruptcy framework. The Dutch bankruptcy law (Faillissementswet, DBA) introduced the procedure «Wet homologate underhand akkoord» based on the preventive restructuring American model, also in Germany, the Government accelerated the adoption of the Directive [12].

In addition, States should encourage the contractual settlement of relations between creditors and the debtor, which is especially important for small companies with concentrated debt structures. In Russian practice, the institution of a settlement agreement has replaced other rehabilitation procedures, due to more «soft» conditions and requirements for its conclusion.

3.2.3 Special (simplified) regimes establishment concerning micro-, small-, and medium-sized enterprises (MSMEs)

Taking into account the fact that in crisis conditions, the bankruptcy highest risks have (MSMEs), one of the trends is characterised by the bankruptcy procedures special (simplified) regimes establishment concerning them. This regime specific is to establish a faster and simplified procedure to minimise costs. For example, new rules for microenterprises have been introduced in Colombia. In Australia, a new insolvency regime for MSMEs came into force on January 1, 2021. The recommendations developed for simplified insolvency procedures rules for small and medium-sized enterprises are the international organisations' responsibility (UNCITRAL, the World Bank) [13]. In jurisdictions where such rules are already in place (USA), attempts are being made to expand the subjects range falling under special treatment, in particular, such rules extension to include medium-sized enterprises [14].

3.2.4 The process of destigmatisation of insolvency

Reducing the stigma of insolvency proceedings is also a policy recommendation suggested by the World Bank and UNCITRAL. The Covid-19 pandemic has caused the spread of the concept that any corporation can meet with a financial crisis. Insolvency is not always due to unfair actions; respectively, business failure could not be punished. This approach contributes to the understanding of insolvency as a tool for business recovery and not only as a liquidation mechanism. Moreover, accordingly, encourages corporations to turn to its mechanisms to get out of the state of crisis.

The most striking example is the change in the «crisis» definition in Italian legislation, which is now defined as a significant economic and financial imbalance. In the legislation of the USA, the term «debtor» is changed than the word «bankrupt» to refer to insolvent companies.

An attempt to reduce the stigma is aimed at analysing the institution of insolvency as an essential part of the entrepreneurial system. An insolvency legislation can actually also for solvent firms. Many studies indicate the need to distinguish between the institutions of insolvency and restructuring, which are components of a single «law of corporate distress» [15]. This approach expands the institutional opportunities of bankruptcy as an anti-crisis regulation tool, amplify it the ability to affect the levels of entrepreneurship and economic growth in a country.

4 Conclusions

The Covid-19 pandemic spread predetermined essentially new challenges for the legislator in the insolvency legal regulation sphere. Responses to the pandemic negative economic consequences in the bankruptcy institution serve as the legal convergence processes a reflection due to the goals and objectives uniformity and the need to identify the most effective model (strategy) to strengthen this institution anti-crisis potential. Thus, despite all the negative consequences of the pandemic, the insolvency institute context can be considered positively, as evidenced by the following arguments. First, the economic crisis phenomena catalysed bankruptcy legislation deficiencies and problems and contributed to its more dynamic reform already existing in the political agenda of many countries. Second, the pandemic facilitated of harmonisation and legal convergence process with undeniably positive effects.

The foreign countries experience in adapting national bankruptcy laws to the pandemic crisis makes it possible to identify the reform most promising areas: improving the effectiveness of rehabilitation procedures, the preventive restructuring mechanisms introduction, special (simplified) regimes establishment concerning micro-, small-, and medium-sized enterprises (MSMEs). Moreover, the economic instability conditions predetermined the insolvency institution concept global revision towards the pro-debitor orientation. The bankruptcy destigmatisation advantages are manifested in the institution liberalisation and increasing its rehabilitation component effectiveness.

References

1. J. Sapir, Economics and Mathematical Methods, **3 (56)**, 5-25 (2020).
2. A. Alemanno, European Journal of Risk Regulation, **11(2)**, 307-316 (2020).
3. A Gurrea-Martínez, *Insolvency Law in Times of COVID-19,* Ibero-American Institute for Law and Finance, Working Paper 2/2020, (available at: https://ssrn.com/abstract=3562685).
4. O.A. Lvova, ECO journal, **6**, 83–102 (2019).
5. J.A. Ellias and G. Triantis, *Congress is ignoring the best solution for troubled companies: bankruptcy,* Fortune, 2020 (available at: https://fortune.com/2020/05/14/bankruptcy-cares-act-aid-coronavirus/).
6. N. Shchukina, V. Varfalovskaya, A. Bekaeva and L. Salinas, Economic Annals- XXI, **1-2 (187)**, 90-98 (2021).
7. A. Jindal, *IBC (Amendment) Ordinance 2020: Straddling Between Manoeuvre and Ambivalence,* IBC Laws Leave a comment, 2020 (available at: https://ibclaw.in).
8. J. Payne, *An Assessment of the UK Restructuring Moratorium*, 2021 (Available at: https://ssrn.com/abstract=3759730).
9. S. Madaus and F.J. Arias, European Company and Financial Law Review, **3-4 (17)**, 318-352 (2020).
10. V.N. Alferov, MIR, **2 (26)**, 180-190 (2016).
11. E. Ghio, J. Gant, G. Boon, D. Ehmke, L. Langkjaer, and E. Vaccari, *Harmonising Insolvency Law in the EU: New Thoughts on Old Ideas in the Wake of the COVID-19 Pandemic.* International Insolvency Law Review, 2021 (Available at: https://ssrn.com/abstract=3855429).
12. S. Madaus, A.K. Wilke and P. Knauth, Int. Corp. Rescue, **17 (1)**, 21-28 (2020).
13. A. Gurrea-Martínez, *Implementing an Insolvency Framework for Micro and Small Firms,* International Insolvency Review, 2021 (Available at: https://ssrn.com/abstract=3715654).
14. W. L. William Norton III & J. B. Bailey, Emory Bankruptcy Dev. J. **383**, 383-393 (2020).
15. S. Madaus, European Business Organization Law Review, **19(3)**, 615-647 (2018).

The Role of Cultural and Creative Industries Sector for Post-COVID Recovery

Gunnar Prause

Wismar University of Applied Sciences, Wismar Business School, 23966 Wismar, Germany

Abstract. Cultural and Creative Industries (CCIs) have been intensively studied within the last decades due to their high economic growth potential and positive impact on innovation. In recent years, research shows increasing efforts to unfold CCIs' potential for innovation partnerships in cross-sectoral cooperation with traditional companies. This classical role of CCI has been discussed in the new context in the frame of COVID-19. Here, the CCI sector itself had to find other channels to meet their clients, but besides that, an extended role of the CCI sector for post-COVID recovery is debated by several key players, including OECD. The paper analyses the current socio-economic situation that is still shadowed by the ongoing COVID pandemic and discusses sustainable and inclusive growth paths of a post-COVID recovery driven by creativity concepts. The Baltic Sea Region (BSR) as the first macro-region with its own macro-regional strategy and its strong innovation power, domination of SME sector, low population density, its heterogeneous business structures resulting from its characteristic mixture of old and new EU Member States can be considered as a test lab for the whole Europe. From the CCI perspective, its long Hanse history that generated a common Baltic identity opens the opportunity to deploy common cultural approaches for successful inter-cultural recovery concepts.
Keywords: Cultural and Creative Industries (CCIs), post-COVID recovery, creativity, The Baltic Sea Region, high economic growth potential.

1 Introduction

The Cultural and Creative Industries (CCIs), or in EU terminology, also referred as to Cultural and Creative Sectors (CCS), enjoyed increasing attention in the last twenty years as a strategic resource for innovation as well as a driving force for entrepreneurial discovery, regional development, and competitiveness [1]. Recent European studies about the CCI sector reveal a contribution in Europe to national GDP between 5-10%, and case studies from different cross-sectoral clusters worldwide highlight the importance of the CCI sector for regional growth and prosperity [2].

Consequently, within the EU Strategy for the Baltic Sea Region (EUSBSR), CCIs have been recognized as an essential contributor to innovation, socio-economic growth, and sustainable and intelligent regional transformation. The Baltic Sea Region (BSR) is the first macro-region with its own macro-regional strategy and placed among the most innovative and advanced Blue Economic regions globally. This fact is surprising since the region is dominated by the SME sector, its heterogeneous business structures, low population, and its mixture of old and new European economies. One important reason for this outstanding performance lies in its long, common Hanse history that coined a common business culture and shaped a Baltic identity facilitating cross-cultural as well as cross-sectoral co-operation and promoting culture to one of the most critical measures for macro-regional development. Hence, cross-cultural similarities pave the road for high-developed cross-sectoral co-operation resulting in successful cooperation of traditional industries with CCIs [3]. Thus, it is no surprise that CCI, together with the "Culture" issue, has found a prominent place on the renewed EUSBSR for the upcoming funding period 2021-2027 of the European Union.

However, this success story of the BSR, together with its CCI sector, is going to be endangered by the COVID pandemic that started in spring 2020 and that affected the cultural and creative sectors along with the tourism sector, most with jobs at risk ranging from 0.8 to 5.5% of employment across OECD regions including BSR [4]. The paper analyses the current socio-economic situation still in the context of the ongoing COVID pandemic and shadows light on tentative sustainable and inclusive growth paths of post-COVID recovery with a focus on the possible role of the CCI sector shortly. The author participated in several EU projects related to the CCI sector in BSR comprising "CTCC project" and "Creative Ports project" so that the presented results are empirically underpinned by studies that took place within these projects between 2017 and 2021 [3].

2 Current socio-economic situation

According to the Eurostat figures, the gross domestic product (GDP) of the European Union declined by about 9% quicker but also sharper during the COVID crisis than it was reported during the financial crisis in 2008/09. This decline came along with growing unemployment figures that have been covered by governmental payment of the seasonal short-time allowance. The initially promised recovery of the economic downturn that was announced for 2021 was postponed due to the ongoing corona pandemic to 2022, and currently, it is not predictable how it will continue. In any case, the level of implications of the COVID crisis to the different EU countries is not comparable since the economically more vital countries in Northern Europe seem to stand the crisis with fewer turbulences than the Southern members. By considering especially the Eurostat figures for Italy and Greece, it has to be stated that both countries did not fully reach the pre-2008 level until the COVID crisis started, and thus they will experience another downgrade of their economic indicators.

Hence, the COVID crisis gave reason for other experts to reflect on the current socio-economic situation. The German economist Andreas Horn coined the term "Zombie Economy" to describe the current economic situation and predict complex post-COVID development [5]. Horn argues that already the Post-financial crisis in 2008/2009 showed already weakness of the economic recovery, and due to political tampering with public money as well as due to bankruptcy protection rules, large parts of the EU economy are kept alive without any economic reason, i.e., too many companies with wrong business models survived and now hindering healthy recovery. In addition, too low interest rates after 2008 took away the necessary financial pressure on companies, and the corona aids of most EU countries delayed the necessary market shakeout.

Horn also studied company investments after 2009 and revealed that too low investment rates reduced the European productivity growth to nearly zero with the consequence that the current lack of workforce is caused by missing gains in labor productivity over the last ten years. This lack of a highly qualified workforce in Europe is further spurred by the social security and tax systems, especially in old EU countries, causing brain drains of talents out of the EU, mainly to Switzerland and USA. Consequently, a. Horn is not very optimistic for post-COVID times by predicting only a short recovery of economy linked with a high number of company bankruptcies, economic stagnation with low growth, and a long-term EU economic downturn. The studies of Horn convinced him that about 15% of all companies represent zombie companies; he expects up to 23% bankruptcies in the year 2022 with a focus on the SME sector with about 2/3 of the cases. Only for Germany this will be around 25 000 companies, and these company bankruptcies represent a severe risk for the banking system that gave credits to enterprises as well as to private persons that are not able to pay back their loans, and they are endangered to fall in private bankruptcy due to unemployment. The situation and risks in other EU countries, especially in Italy and Greece, are significantly higher because these countries are still occupied with recovering from the financial crisis.

Support for Horn's argument comes from the USA, where Wang et al. analyzed company developments in the context of COVID-19, and they estimate the post-COVID bankruptcies in the range of 20-35% starting from 2022 [6]. Comparable figures have been confirmed meanwhile for several European economic institutes. Here, to a significant extent, the tourism sector and real estate, retailing, and CCI sectors are involved. However, this economic and related social decline is additionally accelerated by complementary climate requirements of the European Union where, e.g., in Eastern Estonia, the COVID situation is combined with the decarbonization campaign of the EU confronting the already weak economic base in the region additionally with downsizing requirements of shale oil industry [7].

3 Perspectives

Coming back to the current situation in the CCI sector, it was already stated that the sector was severely hit by COVID pandemia and A. Ricci from EY calculated the loss of revenue in 2020 of European Cultural and Creative Industries to more than 30% leaving thousands of artists with almost no resources [8]. A survey of 2020 conducted within the "Creative Ports project" revealed that the CCI workers are aware of the unstable economic situation related to corona, i.e., most of the surveyed organizations did not plan to close their business or dismiss their employees. Generally, CCI workers are highly qualified, helping them diversify their offered services, especially digitalizing their products. Thus, the majority of the entrepreneurs and CCI employees highlighted the growing opportunity to increase online business activities and pointed out that SMEs are very agile and can adapt quickly to changes because decisions do not request a lengthy approval process. Some CCI actors intensified their promotional activities as well as are searching for new opportunities on the market. The current situation is not a reason to close down their companies as they operate daily. This perception fits the ongoing situation because even though 95% of the museums closed during Corona times, the number of online accesses to online museum websites increased by 200%, revealing that not the cultural demand decreased during corona. However, only the channel the clients consumed cultural services. Unfortunately, the related revenues of all alternative channels of the CCI sector did not keep the standard level [9].

Besides these facts, significant social effects related to the COVID pandemic are discussed since nearly all countries passed lockdown with substantial restrictions related to social contacts to reduce the transfer risk of the virus. Surveys in several countries revealed that these epidemiological measures reduced the number of infections, but they also caused mental problems, depressions, and even suicide due to anxiety, unemployment fears, family aggressions, and social isolation [10]. The German association for depression helped ask about 5000 persons between 18 and 69 years in February 2021 in a representative online panel concerning mental problems related to the COVID situation. The study results showed that the lockdown had a significant negative mental impact on the average population and that 20% of the population that suffered already at least once from depressions 44% indicated a worsening of their mental situation; even 8% had suicidal thoughts or impulses.

These observations motivated OECD to the extent the traditional role of the CCI sector by advocating cross-sectoral co-operation of CCI and health companies since the experiences of lockdowns and social distancing measures made evident the importance of arts and culture for people's mental well-being "and possibly, through the increasingly documented psychosomatic effects of cultural access" also to health [4]. This approach reacts simultaneously to two challenges, first by supporting the CCI economy after the pandemic and second by facilitating and accelerating the recovery of post-COVID mental problems. Such an approach provides new opportunities to capitalize on the role of arts and culture in the prevention and treatment of illness across the lifespan, contributing to solutions for health and welfare systems, such as through reductions in hospitalization or medication rates and can be used to spur digitalization for regional development. The support for artists and firms, which is intended to come from both the public and private sectors, can also leverage culture's economic and social impacts in their broader recovery packages and efforts to transform local economies.

Meanwhile, first cities and regions recognized the advantage of CCI potential for post-COVID recovery and a driver of social-economic impact. Typical areas of CCI involvement in post-COVID recovery are cross-sectoral innovation projects with traditional manufacturing and service companies trying to push besides social and mental recovery issues related to climate change, aging society, or the support of creative and cultural tourism. An essential aspect of the CCI involvement is that the first projects showcase that the concepts are not limited to metropoles, but they also seem to succeed in remote areas. Gerald Vernon-Jackson, Chairman of Culture, Tourism, and Sport of the UK Local Government Association, stressed that "Local creative industries can be the cornerstone of recovery for our communities and local economies from the impact of COVID-19".

Comparable statements are not limited to the UK; they are also expressed in BSR according to the surveys in CCI-related projects comprising "CTCC project" and "Creative Ports project." The open research question for the BSR is to find out which intercultural offers of the CCI sector can facilitate the post-COVID recovery in BSR. Here, more research is necessary to answer this question. The positive message, even with the ongoing COVID pandemic at the end of 2021, is that the outstanding role of the CCI sector becomes more and more hearable in different parts of the world, indicating that the political level gets aware of the social-economic potential of CCI sector to overcome the damaged of COVID pandemic.

4 Conclusions

CCI sector is one of the business branches that have been hit most by the COVID pandemic, but first feedbacks from this sector showed that the CCI workers were not so scared by the economic consequences; they even tried to understand the pandemic as a chance to innovate and find new channels to their clients, especially by using digitalization strategies. Experiences and research concerning cross-sectoral cooperation between CCI and traditional companies delivered encouraging innovation results.

Until now, the mental problems caused by the pandemic, especially related to lockdowns and social isolation, have been neglected in the discussions and psychological studies that highlight severe existing mental problem fields. Thus, post-COVID time requires innovative solutions to overcome these psychological deficits. Here, a promising approach seems to be cross-sectoral cooperation between CCI and the health sector to facilitate and accelerate the post-COVID recovery process. First steps are already taken by initiating culture-health projects for recovery, and the results of these projects seem to be promising.

References

1. J.A. Cunningham, M. Menter, C. O'Kane, R&D Management. **48(1)**, 136–147 (2018)

2. P. Bonetto, B. Hofmann, G. Prause, Journal of Entrepreneurship and Sustainability Issues, **2(1)**, 1−11. DOI: 10.9770/jesi.2014.2(1) (2014)
3. L. Gerlitz, G. Prause, Sustainability, **13** (09742, Economic and Business Aspects of Sustainability), DOI: 10.3390/su13179742 (2021)
4. E. Travkina, P. Sacco, *Culture shock: COVID-19 and the cultural and creative sectors*, OECD (2020) https://www.oecd.org/coronavirus/policy-responses/culture-shock-covid-19-and-the-cultural-and-creative-sectors-08da9e0e/
5. A. Horn, *The Zombie-Economy (in German)*, Edition Novo, ISBN 9783944610610
6. J. Wang, J. Yang, B. Iverson, R. Kluender, *Bankruptcy and the COVID-19 Crisis*, Harvard Business School, Working Paper 21-041, (2020)
7. G. Prause, T. Tuisk, E.O. Olaniyi, Journal of Entrepreneurship and Sustainability Issues, **6 (3)**, 1235−1254. DOI: 10.9770/jesi.2019.6.3(13), (2019)
8. A. Ricci, *A cultural pandemic: How COVID-19 hit Europe's creative sector*, EY (2021)
9. German Depression Barometer, https://www.deutsche-depressionshilfe.de/, (2021)
10. G. Giorgi, L.I. Lecca, F. Alessio, G.L. Finstad, G. Bondanini, L.G. Lulli, G. Arcangeli, N. Mucci, International Journal of Environmental Research and Public Health, **17(21)**, 7857, https://doi.org/10.3390/ijerph17217857, (2020)

Democracy Backsliding as a Danger of Sustainable Development in the Post-Pandemic Period: Social and Legal aspects of Decision

Oleg Tkach[1*], *Oleh* Batrymenko[1], *Dmytro* Nelipa[1], and *Mykola* Khylko[1]

[1]Taras Shevchenko National University of Kyiv, Philosophical Faculty, 01033, 60 Volodymyrska Street, Kyiv, Ukraine

Abstract. The article considers topical issues of the threat of collapse of democracy. Examples of the democracy collapse have shown the lack of free and fair elections in the world, which threatens the independence of the judiciary, restrictions on the right to freedom of speech, which limits the ability of the political opposition to challenge the government, to prosecute, to offer alternatives to the regime. The collapse of democracy in connection with the spread of COVID-19 is being considered, as the democratic spectrum has repeatedly resorted to excessive control, discriminatory restrictions on freedoms such as movement and assembly, and arbitrary or coercive coercion. Attention is drawn to the fact that the outbreak of coronavirus COVID-19 has led to the introduction in all countries of restrictions on the rights and freedoms of the individual in order to prevent the spread of this infectious disease, declared a global pandemic by the World Health Organization. Thus, the unusual nature of the COVID - 19 coronavirus pandemic poses numerous dilemmas to the public, governments, parliaments, the judiciary, law enforcement and many other actors when it comes to the need for effective protection of health and, ultimately, human life, as well as adherence to and ensuring the fundamental democratic principles of man and society. Keywords: democracy, democratic backsliding, political regime, political crisis.

Introduction

The inconsistency of the course of postcommunist transformations with the postulates of the transitological paradigm and its reflection is analyzed in the works of Valerie Bans [1], Juan Linz and Alfred Stepan [2]; the critique of the shortcomings of the transitological paradigm by Thomas Carozers was analyzed [3]. Deviations from democracy are defined in the concept of "double transition" in the work of Adam Przeworski "Democracy and the market. Political and Economic Reforms in Eastern Europe and Latin America" [4], the concept of "triple transition" in the work of Klaus Offe "Capitalism through a democratic project? The Democratic Theory of the Triple Transition in Central and Eastern Europe "[5], the concept of the "Quarter Transition" in Taras Kuzio's "Transition in the Post-Communist States: Triple or Quadruple" [6], the concept of the "fourth wave" of Michael

* Corresponding author: tio19@ukr.net

McFaul's political transitions in the Fourth the wave of democracy and dictatorship: uncoordinated transitions in the postcommunist world" [7], in the hypothesis of the third reverse wave.

It is undeniable that the effects and long-term consequences of the COVID-19 pandemic are still unclear. One thing is sure: the impact (mostly negative) of the coronavirus on human rights and freedoms is highly significant.

In this aspect, the socio-cultural dynamics of the world, which in current conditions has taken the form of social and legal turbulence, highlights the problem of finding a political organization that would suit the majority of the population, corresponds to fundamental democratic principles, not violates them. On the other hand, States are expected to anticipate appropriate threats, be proactive, and take any appropriate action in advance, applying the so-called "prevention principle".

The application of quarantine restrictions in the aspect of counteracting the spread of coronavirus has repeatedly attracted the attention of scientists. It does not seem necessary, possible and expedient to list everyone's surnames. However, we highlight those whose works are exciting in the study of the issue. Thus, in particular N.O. Vasyuk studied the role and powers of local public authorities in ensuring the sanitary and epidemic well-being of the population [8];

Regarding the problems of observance and restriction of fundamental human rights in the context of preventing the spread of COVID-19, Joseph J. Amon (2020) studied the issue of preventing the spread of COVID-19 among prisoners in the context of ensuring their right to medical care [9];

In terms of research, A.S. Frost's article "Pandemic COVID-19: Threats and Opportunities for Democracy" [10].

The monograph of O.V. Banchuk also seems to be scientifically valuable. "International practice of quarantine restrictions on the rights and freedoms of the individual", in which she analyzes in detail the international practice of quarantine restrictions on the rights and freedoms of the individual [11].

However, despite the diversity of views, there are currently no comprehensive and comprehensive studies that would allow us to determine the specifics of this phenomenon's impact on democracy. It should be noted that ensuring the proper implementation of fundamental human rights during the COVID-19 pandemic requires a more in-depth analysis of the theoretical provisions and legal framework for establishing and implementing guarantees and restrictions on civil and political human rights in preventing the spread of COVID-19 in Ukraine. And the shortcomings of the anti-epidemic measures introduced in this regard, identifying the causes, conditions, and specifics of the destruction of democracy in a pandemic and finding ways to reduce the negative impact of the pandemic on democratic foundations.

Thus, all of the above makes the study extremely relevant. Most studies of the state of modern democracy begin with the thesis that after the end of the Cold War, it seemed that democracy was spreading. The collapse of democracy can take place in several ways: because the process is often led by democratically elected leaders who use "gradual rather than revolutionary tactics"; the process of return is manifested through "slow, barely noticeable steps"; through covert authoritarianism to describe the practice of an authoritarian leader (or potential authoritarian leader) who uses "imaginary legitimate legal mechanisms for anti-democratic purposes, concealing anti-democratic practices under the guise of law"; rejection or insufficient level of adherence to democratic rules of the game, denial of the legitimacy of political opponents, tolerance or encouragement, willingness to restrict the freedoms of opponents, including the media; manipulation of defamation laws, election laws or laws on "terrorism" as tools for persecuting and discrediting political opponents; use of democratic rhetoric as a distraction from anti-democratic practices, as a manifestation of hidden authoritarianism; through authoritarian political culture as a factor contributing to the retreat from democracy [5].

Methodology

Due to the nature of the topic, various methods were used. Data on the effects of the pandemic on democracy were collected and systematized using the induction method. Based on these collected data, the results and recommendations were developed using the deduction method. Thus, in particular, the systemic approach provided an opportunity to consider the phenomenon of democracy as a whole and concerning the pandemic. It is due to a systematic approach that the relationship between the negative impact of the pandemic on the democratic foundations of the state and society becomes logical. The institutional approach is used to study the structures through which the state's activities are aimed at protecting the population from the adverse effects of the pandemic. The structural-functional approach made it possible to consider anticoronavirus policy as unquestionable integrity, a system with a complex structure, each element of which has a specific purpose and performs specific functions aimed at combating threatening manifestations and trends. The sociological approach aimed to determine the socio-political factors of the pandemic, the dependence of the functioning of democracy on the impact of the pandemic, the conditionality of political processes, as well as the impact on democracy of other processes caused by the pandemic.

The aim of the study is to determine the causes of the collapse of democracy in modern conditions, to analyze the causal links of the erosion of democracy under the influence of the pandemic, as well as to find possible ways to counteract this negative phenomenon. Democracy backsliding as a danger of sustainable gevelopment in the Post-Pandemic Period: social and legal aspects of decision.

3 Results and Discussion

At the same time, these scientific works mostly concern only general issues of protecting the population from epidemic diseases, without considering the current legal regulations and the practice of preventing the spread of COVID-19 in Ukraine. And the problem of destroying the foundations of democracy in terms of preventing the spread of COVID-19 has not yet received adequate coverage in domestic and foreign scientific literature, given the relatively recent emergence and ongoing pandemic, as well as the constant review of measures against Ukraine and other countries.

The use by most countries of restrictions on civil rights and freedoms at the level of martial law or a state of emergency has a complex dialectical nature. On the one hand, quarantine measures are taken for the benefit and benefit of humans - as the most effective means to reduce the occurrence of severe acute respiratory syndrome coronavirus 2 (SARS-CoV-2), which causes COVID-19 in a significant number of people.

On the other hand, quarantine severely restricts people's rights and freedoms, restricts their ability to work, earn an income and meet their basic needs. According to a group of American and British scholars, Thieme Fetzer, Lucas Hensel, Jones Herml, and Christopher Roose, the measures taken "affect civil liberties, erode social capital, and cause economic insecurity."

Domestic and foreign scholars have paid attention to problematic legal issues of national security in the field of health care and the protection of fundamental rights and freedoms of citizens. Yes, V.M. Pashkov studied the legal characteristics and classification of threats in health care from the standpoint of commercial law [13]. V.V. Borsch considered the national health care system as a structural element of the national security of Ukraine [14]. G.I. Rybak studied the improvement of the population's quality of life as a factor in ensuring national security. However, Ringach drew attention to public health as a factor of national security [15]. M. Feinberg, L. Niada-Avshalom and B. Toebes studied human rights in national security and public health [16]. W. Golder and G. Williams drew attention to human rights and national security in the fight against terrorism [17].

"How will the coronavirus change democracy and governance around the world?" Under this name, on April 6, 2020, the Carnegie Endowment for International Peace published its forecast. In their study, authors Frances Z. Brown, Saskia Brechenmacher and Thomas Carothers give a generally negative outlook for the development of democracy around the world. Experts note that "the pandemic arose at a time when democracy was already in danger in many places, which in turn threatens to increase the backlog of democratization and authoritarian consolidation."

At the same time, the authors believe that in some cases, the effective response of the state can strengthen confidence in the government, and the inability to hold elections can lead to the necessary innovations in their administration.

Similar to the Carnegie Endowment for International Peace, COVID-19's assessment of democracy's impact is provided by Spanish scholar Francis Amat et al. In their study, the researchers describe the results of a survey conducted on March 20-28, 2020 in Spain. These results indicate an increase in the widespread demand of Spaniards for solid leadership, a willingness to renounce individual rights and freedoms and a sharp increase in support for technocratic and authoritarian rule [19, p. 24].

Isolation of previously unresolved issues. Despite intensive democratization at the turn of the century, the role of democracy has not become leading. Therefore, research needs such aspects as the influence of civil society on the preservation of democratic values, institutional transformation, increasing the influence of innovation forces.

The problem of the collapse of democracy has a special place, because there are theoretical and methodological challenges: first, despite the diverse literature, there is a lack of theories to explain the deviation, and secondly, there are theoretical debates, for example, on the causes of democratic transition, democracy consolidations that remain unresolved. The methodological challenge involves measuring regime change. Democratic decline is accompanied by the weakening of political institutions, the process of peaceful transfer of power, changes in important components of democracy, such as violations of rights and freedoms.

Researcher Frances Z. Brown highlights a number of negative consequences of the coronavirus pandemic for democracy, namely: Centralization of power. Illiberal leaders have the opportunity to take advantage of the crisis to weaken checks and balances and destroy accountability mechanisms, thereby consolidating their power. In Hungary, for example, a new law allows Prime Minister Viktor Orbán to run the government indefinitely, without any parliamentary scrutiny. Restrictions on fundamental human rights. One of the clear trends is increased control over the media under the guise of combating "misinformation" about the virus. The Egyptian government recently forced a reporter from The Guardian to leave the country after asking about the official number of coronavirus cases in Egypt. Strengthened state supervision. In Israel and South Korea, governments use the location of smartphones to track citizens who could potentially be carriers of the virus. In Hong Kong, newcomers must wear electronic bracelets to track the whereabouts of individuals. Restriction of protests. There is a risk that governments may use the current need to restrict public gatherings as an excuse to combat the wave of anti-government protests that have swept global politics over the past few years. For example, in Algeria, where last year's protests pushed the government for some political reforms, the government banned all protests, marches and demonstrations. "Interruptions" in elections.

The pandemic threatens to disrupt electoral processes around the world. The United States has already delayed several early state-level voting, and candidates have stopped campaigning. Several European countries, including Italy, northern Macedonia, Serbia, Spain and the United Kingdom, have postponed national or local elections. Imbalance of civil-military relations. Responding to a crisis can change the balance of power between military and civilian authorities. In many countries, from Iran and South Africa to Israel and Peru, the military has been called in to help counter the pandemic by stopping people and in other ways. Pressure on civil society. The government's introduction of pandemic emergencies is exacerbating the already significant trend of shrinking space for civil society in many parts of the world. Extreme restrictions on movement, information gathering and confidentiality do not work in favour of civil society organizations. Resumption of the discussion "authoritarianism against democracy". The pandemic will put strong pressure not only on government agencies in the most affected countries, especially health care systems, but also on many other important government functions, from education and food supply to law enforcement and border control. The acute public health situation, coupled with insufficient government capacity, weak institutions, limited access to information and low public confidence in leaders (and consequent reluctance to adhere to health directives), may resume the debate on the benefits of authoritarian approaches to crisis management.

In contrast, a survey of Western European public opinion by French researcher Andre Blair points to some positive developments in citizens' attitudes toward democratic institutions and procedures. Researchers point out that people understand the need for these severe social restrictions and are willing to follow the will of the governments that have decided to apply them, at least in the short term. Scholars attribute this fact to the probability of citizens' awareness of such difficult government decisions, the priority of which is the health of vulnerable people, rather than economic interests [19, p. 10-11].

In general, agreeing with the researchers' conclusions on the impact of the COVID-19 coronavirus pandemic on the functioning of democratic institutions, we must note that the pandemic is an emergency situation that requires maximum mobilization of resources of states, international organizations and the public to save lives and health, which is the first task of democracy. The restrictive measures taken are inherently contrary to key democratic ideals, such as freedom, human dignity, the right to privacy, and violate the fundamental principle of the rule of law.

It is safe to say that the measures taken pose a potential threat to typical democratic procedures, as well as to the rights and freedoms of citizens. However, whether these threats will become a reality, whether they will be applied in a discriminatory manner, whether the government will become more authoritarian and people less free, whether there will be more censorship and less civil rights depends on many factors. We believe that the critical issue to be monitored in the near future is how bans will be lifted in the quarantine easing process or whether government officials will try to keep them indefinitely, with the opportunity to use them to their advantage.

Along with the possible threats, we believe that the pandemic will create several positive opportunities for democracy, as paradoxical as it may sound. In particular, objective preconditions will force the state, the expert community, and the public to accelerate the introduction of innovation processes in democratic procedures and institutions.

In our opinion, the pandemic can give impetus to a new round of development of e-democracy institutions. By analogy with business, education, finance, which is a pandemic implement remote forms of work and provide remote access to customers to their services, the sphere of public policy will undergo similar transformations. Accordingly, governments must learn to communicate with each other and with citizens using modern information and communication technologies without losing efficiency and without increasing costs. Thus, the COVID-19 coronavirus pandemic should provide an impetus for innovation in electoral technologies and voting processes, public access to administrative services, e-government and other democratic institutions.

Yascha Mounk, an adjunct professor at Johns Hopkins University and a senior fellow at the Council on Foreign Relations, described Democracy on the Defense as optimistic about the spread of democracy worldwide. However, this confident optimism was inappropriate. Looking back, it is clear that it was naive to expect democracy to spread to all parts of the world. The authoritarian turn of recent years reflects the shortcomings of democratic systems. But they omit an important part of the picture. The history of the past two decades - is not just a story of the weakness of democracy; it also determines authoritarian capacity. According to official statements, the democratization of the world is proclaimed one of the main goals of U.S. foreign policy, but these statements are seriously criticized. This direction of foreign policy is due to the popular theory of the democratic world in the United States, which is mentioned in the speeches of leaders of both dominant parties. In the above-mentioned article Yasha Munch defined that the United States seeks to restore the role of "leader of the free world." Biden and his team have announced their intention to convene a summit of democracies to support democracies in combating the threats of autocrats, which will signal the U.S. commitment to democratic values. The problem is that the two main goals of the effort (deterring the influence of powerful autocracies and preventing collapse in key democracies) often contradict each other [20].

Therefore, the American researcher J. Munch determines that in the coming decades there should be a long struggle between democracy and autocracy. Since the 1990s, autocratic regimes have advanced in terms of economic performance and military power. Leaders have learned to use digital tools to subtly suppress opposition movements. They repulsed democratic campaigns, significantly increased international influence, this is not only a retreat from democracy, but a revival of authoritarian regimes [20].

The result of this competition has not been determined. To win, the United States and its democratic allies need to understand the importance of these historic moments, working together to defend global democracy in creative, bold ways, unlike in the past. They will also need to resolve the dilemma arising from the contradictions between the two main goals: on the one hand, to stop retreats in their ranks, on the other hand, to support a united front against authoritarian regimes such as those in China and Russia. Addressing this dilemma requires a skilful approach that preserves the possibility of cooperating with countries that have questionable democratic principles of good faith while maintaining a close partnership for indeed democratic allies. It will also mean abandoning the "promotion of democracy" in favour of "protecting democracy" for security purposes. Yasha Munch stressed that the step in preventing the revival of the authoritarian regime is the reform of two prominent institutions of the liberal international order: the E.U. and NATO. The Americans and Europeans who created these bodies believed that their own countries would never have the collapse of democracy. As a result of this approach, these organizations do not have easy means to suspend membership or expel a member whose behaviour has changed. This is particularly problematic for the E.U., which requires its members to sacrifice a degree of sovereignty to join the bloc. It is sometimes complicated for national politicians to explain this to their constituents. Most E.U. countries do not address transnational issues, such as climate change or the impact on world politics. Because these countries uphold democracy, the rule of law, and the renunciation of a degree of independence, allowing them to promote common democratic values. However, the increase in the number of authoritarian leaders in the E.U. (Hungary, etc.) undermines the legitimacy of the bloc. It is crucial to resolve this contradiction, and the E.U. is facing a legitimacy crisis that current institutions cannot resolve. This requires political capital, diplomatic pressure, legal, organizational rethinking.

In an article on the collapse of democracy in 2018, researchers L. Janetti (Licia Cianetti), James Dawson and Seán Hanley argued that the emergence of populist movements in Central and Eastern Europe, such as "Babiš ANO" in the Czech Republic, is "Potentially ambiguous phenomenon, formulating real social needs in political reform, putting the issue of effective governance in the spotlight" [21].

In a 2019 document presented by the International Association of Political Psychologists, Sean Rosenberg argues that right-wing populism exposes the vulnerability of democratic structures and that "democracy is likely to consume itself."

Researchers are determining the impact of the COVID-19 pandemic on the collapse of democracy. Many national governments have postponed or cancelled democratic elections at both the national and subnational levels due to the COVID-19 pandemic, which has identified gaps in democratic practices.

Thus, the local elections in the U.K. in 2020 were postponed for a year. At the beginning of the crisis, U.N. experts recommended appropriate measures to the government, which were "necessary and non-discriminatory" [24].

The collapse of democracy through a pandemic is a unique deviation from democracy associated with national crises. This happens when leaders impose authoritarian rules during a state of emergency, the crisis in the country, or remain in place after the situation improves. This happened in several countries during the coronavirus pandemic.

According to the V-Dem Institute, only 39% of all countries did not commit minor violations of democratic standards in response to Covid-19 [25].

Even though liberal democracy was defended and experienced an increase in the number of autocrats and authoritarian regimes in many parts of the world until the first death from the coronavirus in December 2019, the pandemic affected the collapse of democracy [26].

Given that COVID-19 spread worldwide in early 2020, the global decline in freedom has intensified [27].

The outbreak exposed the weaknesses of all components of democracy, from elections and the rule of law to extremely disproportionate restrictions on freedom of assembly and movement. Democracies and autocracies have had successes and failures in the fight against the virus itself, although citizens in authoritarian states have lacked the tools to confront them. After all, the changes caused by the pandemic have left many societies (with different regimes, incomes and demographics) in worse political situations, with more pronounced racial, ethnic and gender characteristics of inequality, vulnerable to long-term consequences. Transparency has been one of the most affected aspects of democratic governance. National and local officials in China have diligently blocked information about the outbreak, including through mass arrests of Internet users who shared the information.

In December, journalist Zhang Zhan was sentenced to four years in prison for reporting Wuhan as the initial epicenter. The Belarusian government has been actively downplaying the seriousness of the pandemic to the public, which has refused to take action, while the Iranian regime has concealed the real victims of the virus. Some repressive governments, including Turkmenistan and Nicaragua, have ignored reality and denied the presence of the pathogen in their territory. More open political systems have also experienced significant problems with transparent information.

At the presidential, state level, U.S. officials provided fictitious data and actively spread misinformation about the transmission and treatment of coronavirus, leading to confusion and politicization of what was to become a health issue. Similarly, Brazilian President Jair Bolsonaro has repeatedly promoted untested treatments that have cast doubt on the usefulness of masks and vaccines.

Governments have also deployed surveillance measures, which have often been of dubious value to public health and have a limited number of precautionary measures against abuse. However, beyond their effects, in 2020, official responses to COVID-19 laid the groundwork for government excesses that could affect democracy. As in the aftermath of the September 11, 2001, terrorist attacks, as the United States and many other countries dramatically expanded their oversight and restriction of the right to national security, the COVID-19 pandemic led to changes in regulations and problematic legislation that would be difficult to repeal once overcome. Virus. For example, emergency measures were taken in Hungary, which allowed the government to issue decrees, even though coronavirus cases were insignificant. Among other abuses of these new powers, the government withdrew financial aid from municipalities led by opposition parties. Motivation for better executive power and increasing the level of authority contributed to the gradual concentration of power of Prime Minister Viktor Orban [26].

In Algeria, President Abdelmadjid Tebboune took office due to tightly controlled elections after the authoritarian leader Abdelaziz Bouteflika, who resigned under public pressure, banned all mass activity and assembly. Even when other restrictions were eased in June, the ban on rallies remained, authorities stepped up arrests of pro-democracy activists. Many of the arrests were based on the April amendments to the Criminal Code. The amended code increased convictions for defamation and criminalized the dissemination of false information.

Indonesia has approached the military and other security forces as crucial players in the pandemic. Much military personnel have been assigned to leadership positions in the country as a COVID-19 task force, and the military has provided significant support for the development of ambulance hospitals and the provision of medical supply services.

The collapse of democracy occurs when the components of democracy are threatened, due to the lack of free and fair elections, rights to freedom of speech, associations; bureaucratic constraints of the government, for example, when there is a threat to the independence of the judiciary, which limits the ability of the political opposition to challenge the government, to demand responsibility, to offer alternatives to the regime.

A 2020 study using the World Values Survey found that proponents of cultural conservatism were the ideological group most open to authoritarian rule in Western democracies. In English-speaking Western democracies, which are "based on the protection" of relations, cultural conservatism and left-wing economic views, which supported authoritarian regimes, are combined.

Stephen Haggard and Robert Kaufmann identify three critical reasons for the collapse of democracy: "the effects of polarization; restructuring of party systems, which allows elected autocrats to gain legislative power; the gradual nature of the retreats, which divides the opposition, throws off-balance".

The COVID-19 crisis is an alarm signal, an urgent warning that freedoms are under threat. Democracy is now under threat, and people who care about it must show the will, discipline and solidarity to protect it. The freedom, health and dignity of people around the world are at stake. The essence of the public interest in a pandemic is the need to save people's lives and restore their health. In view of this, the legislature has the right to establish rules of conduct for all subjects of law, including those that restrict human rights and freedoms.

Democracy has suffered new losses in the fight against authoritarian supporters, shifting the international balance in favor of authoritarianism. Incumbent leaders are more likely to use force to remove opponents and settle accounts, ostensibly for health care, while activists, lacking effective international support, have faced severe sentences, torture or assassination in various circumstances, and a decline in global freedom. Waves of false information were generated. There was a misconception that democracy was declining because it could not solve people's problems and needs.

Prospects for further research of the scientific problem: a number of problems related to the process of obtaining administrative services by citizens remain unresolved, which should be addressed not only at the local but also at the national level. Therefore, the elimination of the above problems requires a comprehensive approach, namely: the adoption of new regulations and improve existing ones in order to simplify the procedure for obtaining administrative services; creation of remote workplaces and mobile offices in order to ensure the availability of administrative services for the population not only in large settlements, but also throughout Ukraine.

Conclusions

Democracy is not just a cherished ideal but a phenomenon that is the public administration system best suited to overcome the crisis of scale and complexity of COVID-19. Only through democracy can societies strengthen the social trust that allows them to continue to live in a crisis, maintain national resilience in the face of difficulties, resolve profound differences in society through full participation and dialogue, and maintain confidence that the return will be shared and the rights of all citizens will be respected and guaranteed.

The impact of the COVID-19 coronavirus pandemic on the functioning of democratic institutions shows that the pandemic is an emergency that requires maximum mobilization of state resources. A proper and effective anti-crisis policy of the state is essential for overcoming or at least reducing the negative impact of the pandemic on democracy. No less important is that the Ukrainian government must also actively care about security - much more than in those countries where there is no war. Especially since many citizens are vigilant to ensure that it does not make significant concessions to the enemy, and in a pandemic, their vigilance has only increased. In contrast to the selfish allegations of authoritarian propaganda, credible and information-based flows of strategic options, voluntary self-organization of civil society, and open interaction between government and society are all vital tools in the fight against a pandemic. Only with the help of democracy can an independent civil society be allowed to establish partnerships with state institutions. Only through democracy can society ensure a sustainable balance between competing needs and priorities - between fighting the spread of the virus and protecting people's civil and political rights in accordance with constitutional guarantees.

The consequences of the coronavirus pandemic's impact on democracy are centralization of power; restriction of fundamental human rights; enhanced state supervision; restriction of protests; "Interruptions" in elections; imbalance of civil-military relations; pressure on civil society; resumption of the discussion "authoritarianism against democracy". The pandemic crisis has stimulated the demand for the involvement and use of digital technologies in various spheres of society and the state.

Pandemic Crisis management involves assessing the appropriate duration of the quarantine regime, taking into account all identified risks and ensuring appropriate adaptation of crisis measures, in particular: creating a multilevel risk management system and improving strategic planning and analysis as essential elements of crisis management and decision-making. Rights and freedoms of citizens; providing society with the necessary support to prevent the irreversible negative consequences of a pandemic, etc. With the spread of COVID-19, the democratic spectrum has repeatedly resorted to excessive control, discriminatory restrictions on freedoms, such as movement and mass assembly, and arbitrary or coercive coercion to such restrictions.

References

1. V. Bunce, World Politics, **55(2)**, 167-192 (2003).
2. J. Linz. *Problems of Democratic Transition and Consolidation: Southern Europe, South America, and Post-Communist Europe* (Johns Hopkins University Press, 26, 1996).
3. Th. Carothers, Journal of Democracy, **13(1)**, 5–21 (2002).
4. A. Przeworski, *Democracy and the Market: Political and Economic Reforms in Eastern Europe and Latin America* (Cambridge, 1991). https://doi.org/10.1017/CBO9781139172493
5. C. Offe, Social Research, **71**(3) (2004).
6. T, Kuzio, Politics, **1(3)**, 168–177 (2001).
7. M. McFaul, World Politics, **54 (2)**, 212–244 (2002).
8. Н. О. Васюк, Інвестиції: практика та досвід, **20**, 89–93 (2010).
9. J. Amon, Health and Human Rights Journal. URL: https://www.hhrjournal.org/ 2020/03 / covid-19-and-detention-respecting-human-rights.
10. А.С. Мороз, *Пандемія COVID-19: загрози та можливості для демократії.* URL: https://dspace.nau.edu.ua/bitstream/NAU/43315/1/2020.05.01%20CO VID-19.and.democracy.Moroz.pdf
11. Банчук-Петросова О.В. *Міжнародна практика карантинних обмежень прав та свобод особи*. URL: http://pravoisuspilstvo.org.ua/archive/2020/ 5_2020/36.pdf
12. T. Fetzer, L. Hensel, J. Hermle, Ch. Roth, arXiv Working Paper. URL: https://arxiv.org/abs/2003.03848 (2020).
13. В.М. Пашков, Підприємництво, господарство і право, **12**, 134–137 (2007).
14. V.V. Borshch, Науковий вісник Ужгородського національного університету. Серія «Міжнародні економічні відносини та світове господарство», **23(1)**, 19–23 (2019).
15. H.I. Rybak, Потенціал економіки країни: сучасний стан, розвиток та управління, **6,** 44–45 (2016).
16. M. Feinberg, L. Toebes, The International Journal of Human Rights, **19(4)**, 383–387 (2015).
17. B. Golder, G.Williams, Journal of Comparative Policy Analysis, **8 (01)**, 43–62 (2006).
18. F. Amat, A. Arenas, Albert Falcó-Gimeno, Jordi Muñoz, *Pandemics meet democracy: Experimental evidence from the COVID-19 crisis in Spain* (University of Barcelona, 2020) URL: https://osf.io/preprints/ socarxiv / dkusw / (2020).
19. Y. Mounk, Democracy on the Defense. Foreign Affairs (2021). URL: www.foreignaffairs.com/articles/united-states/2021-02-16/democracy-defense
20. L. Cianetti; J. Dawson, East European Politics, **34 (3)**, 243–256 (2018).
21. S. Rosenberg. *Democracy Devouring Itself: The Rise of the Incompetent Citizen and the Appeal of Right-Wing Populism* (eScholarship, University of California, 2019).
22. *Democracy under Lockdown*, in Freedom House, 28 January (2021).
23. *English local elections postponed over coronavirus*, in BBC News, URL: https: // www . v-dem.net. 13 March (2020).
24. *Pandemic Backsliding V-Dem*, URL:https: // www . v-dem.net. (2021).
25. OHCHR. COVID-19. 2021: States should not abuse emergency measures to suppress human rights - UN experts. - Mode of access: https://www.ohchr.org/EN/NewsEvents/Pages/DisplayNews.aspx?NewsID=2572

26. Democracy under Siege. Freedom in. the world 2021. Highlights from Freedom House's annual report on political rights and civil liberties, ttps: // www . freedomhouse.org/sites/default/files/2021-02/FIW2021_World_02252021_FINAL-web-upload.pdf. (2021).
27. World Values Survey. WVS Database, https: // www. Worldvaluessurvey. (2020).

The Impact of COVID-19 on the Development of the Segment of the International Tourism Market in Ukraine Focused on Arab Tourists

Sergii Illiashenko[1], *Nataliia* Illiashenko[2,3], *Yuliia* Shypulina[4], *Ievgeniia* Golysheva[3]*, *Natalia* Tomczewska-Popowycz[5], *Małgorzata* Rutkowska[6], and *Janusz* Klisinski[7]

[1]University of Economics and Humanities, Department of Marketing, 43-300, ul. gen. Wł. Sikorskiego 4-4c, Bielsko-Biala, Poland
[2]Sumy State Pedagogical University named after A. S. Makarenka, Department of Business Economics and Administration, 40002, Romenska str., 87, Sumy, Ukraine
[3]Sumy National Agricultural University, Research Section, 40021, Herasyma Kondratieva str., 160, Sumy, Ukraine
[4]National Technical University "Kharkiv Polytechnic Institute", Department of Economic Cybernetics and Marketing Management, 61002, Kyrpychova str., 2, Kharkiv, Ukraine
[5]University of Silesia in Katowice, Institute of Socio-Economic Geography and Spatial Management, 41200, Bedzinska str. 60, Sosnowiec, Poland
[6]Wroclaw University of Science and Technology, Department of Organization and Management, 50370, Wybrzeże Wyspiańskiego str. 27, Wroclaw, Poland
[7]Technical-Humanistic Academy, Department of Marketing and Entrepreneurship, 43309, Willowa str. 2, Bielsko-Biala, Poland

Abstract. The article is devoted to determining the set of external and internal prerequisites for the extending influx of Arab countries' tourists to Ukraine in 2021 under the COVID-19 pandemic restrictions introduced in most countries. Quantitative characteristics of the flow of tourists are outlined. The main characteristics and preferences of tourists from Arab countries who visited Ukraine are determined. It is established that in conditions of restrictions caused by COVID-19, they consider Ukraine as an alternative vocation location to traditional places in European countries. A generalized profile of these tourists has been compiled. It can serve as an information base for improving tourist products (services) and the system of service rendering. An assessment of prospects for developing the international tourism market segment in Ukraine focused on tourists from Arab countries has been made. The study results can be used to make effective management decisions in the field of international tourism at the state and regional levels, as well as at the level of individual tourism service providers in the context of the continuation of the COVID-19 pandemic and its completion. Keywords: international tourism, COVID-19 pandemic, Ukrainian tourist market, development prospects, Arab tourists.

1 Introduction

* Corresponding author: golysheva@gmail.com

The COVID-19 pandemic has negatively affected the entire world economy, especially the tourism industry. Over a hundred countries have imposed various travel restrictions. As a result, revenues from the tourism industry have decreased significantly. According to [1], the average share of tourism in the economy of foreign countries is about 10%, and it suffered significant losses. At the same time, the share of foreign tourists is up to 90% of the total industry in countries with a developed tourism sphere. Ukraine's tourism industry is less dependent on foreign tourists, accounting for about one-third of the total. However, the situation in Ukraine is somewhat paradoxical - the COVID-19 restrictions, introduced in most countries of the world, facilitated the influx of foreign tourists to Ukraine in 2021, primarily from Arab countries. According to the preliminary analysis results, tourists from Arab countries, who previously gave preferences to other European countries, began visiting Ukraine. These are tourists from Saudi Arabia, the United Arab Emirates, Kuwait, and other countries. According to tour operators' information, the majority of tourists are wealthy people making family vacation trips. This will significantly increase the income of the tourism industry. According to the "Join UP!" travel agency [2, 3], the tourism industry of Ukraine will make an additional UAH 1 billion from tourists from Saudi Arabia and Kuwait in 2021. Consequently, it is necessary to assess the prospects of consolidating the consumer preferences of existing groups of Arab tourists and encouraging their influx to Ukraine. This will purposefully form a system of activity to stimulate the development of the international tourism market in Ukraine both during the COVID-19 pandemic and after its overcoming.

2 International tourism in a COVID-19 pandemic: literature review

The impact of the COVID-19 pandemic on tourism has been studied in Abbas et al. [4]. Its authors note significant losses from the pandemic and state the need to restore tourism and traveling activity. In this context, an analysis of methods to restore the tourism industry has been produced. Recommendations on the development and implementation of activities to restore the tourism sector are offered to tourist organizations, public authorities, scientists, etc.

Uğur, Akbıyık [5] investigated the reaction of travelers to the COVID-19 pandemic using text mining methods in online travel forums. The results of in-depth analysis of the texts revealed the impact of travel insurance on decision-making by travelers. The authors consider travel insurance as a method of reviving the tourism industry during a pandemic. It shows the negative impact of the pandemic on tourists themselves and the tourism industry as a whole.

In Saha et al. [6], the impact of the COVID-19 pandemic on the tourism sector of regions and countries was studied. On this basis, a countries ranking was built, which considers the impact of tourism on their economies. The impact of the COVID-19 pandemic on tourism and the world economy has been assessed.

Aronica et al. [7] performed a forecast of the future scenario of the tourism industry based on retrospective data. It provides a steady reduction in foreign tourists, with the most significant reductions expected in developing countries. The main macro-factors that will affect the consequences of the COVID-19 pandemic in different countries are outlined.

Neumayer et al. [8] analyzed a sample of bilateral travel restrictions between 27 European countries for six months during the second wave of the pandemic. Based on the received data, three forecasts made according to epidemiological, economic, and political logic are checked. The results of the analysis show that economic and political logic strongly influences travel restriction. This confirms the need for European countries to pursue a coherent policy of restrictions.

Gegung [9] analyzes the possibility of using virtual travel to traditional tourist locations during the COVID-19 pandemic and after it to maintain sustainable attendance. It also contributes to the ecological, cultural, and economically sustainable development of tourist locations and reduces the risk of disease.

Cretu et al. [10] investigated the tourists' attitude to traveling during the COVID-19 pandemic. The study was conducted using the online survey of the active population of Romania. The survey period coincided with the third wave of the pandemic. The results showed that fear of infection and risk significantly affected the intention to travel. These intentions depend on the state of morbidity and protection conditions in the places of travel. However, the very intentions to travel for vacation or leisure have survived.

Kitamura et al. [11] assessed changes in the tourism industry, carbon emissions, and employment in Japan due to the COVID-19 pandemic. They identified and analyzed four scenarios of these changes, and their assessment was performed. Recommendations for overcoming negative economic and social changes and maintaining positive environmental changes in the post-pandemic period were developed.

Tsai [12] is exploring effective strategies for developing Taiwan's tourism industry, which focuses on the most promising consumers of tourism services. The main focus is on analyzing statistics characterizing tourists from China, Japan, and Southeast Asia visits to Taiwan. Treatment and/or vocation tourist directions are analyzed. Recommendations for the formation of strategies for sustainable development of tourism have been developed.

Binh Do et al. [13] investigated the impact of the COVID-19 pandemic on Vietnamese tour operators and analyzed their activity to overcome the predicament. The research is based on the results of a survey initiated by the National Tourism Advisory Council and in-depth interviews with 32 heads of Vietnamese travel companies. Studies have shown a sharp deterioration of the tour operators' business. At the same time, they implement short-term and medium-term anti-crisis activities and practically do not contemplate long-term strategies. The paper emphasizes implementing long-term strategies for the tourism industry to emerge from the crisis caused by the pandemic.

Rahman et al. [14] analyzed the impact of the COVID-19 pandemic on the travel risks of tourists and their perception by tourism business leaders. The initial data were collected through social networks; the sample includes 716 respondents. The obtained results are intended to increase the validity of tourism travel risk management caused by COVID-19.

Hamid et al. [15] conducted a study of the impact of COVID-19 on the tourism industry and hotel services worldwide and in detail (including impact on economic contractors in the field of tourism) in Malaysia. The issue of social integration due to the pandemically accelerated introduction of new creative forms of tourism, which are focused on market niches, was studied. It was concluded that, unlike previous crises, a rapid economic and tourism recovery after the COVID-19 crisis is doubtful. Prokopenko, Omelyanenko [16], on the example of the Baltic region, studied issues of internationalization of clusters of cultural and creative industry, including tourism.

Cheng, Zetina [17] investigated the effects of COVID-19 on the tourism sector in Belize. They showed that the pandemic caused a crisis in the tourism sector and the country's economy as a whole, and it takes time and significant funds to overcome it. Recommendations for overcoming the global crisis in the field of tourism caused by the COVID-19 pandemic were offered.

Pashkus et al. [18] studied the impact of the COVID-19 pandemic on the tourist competitiveness of St. Petersburg (Russia). An approach to determining the competitive advantages of attractive areas for tourism in the post-pandemic period was proposed.

Tomczewska - Popowycz, Quirini - Popławski [19] studied the impact of political instability in Ukraine in the last five years on tourist flows to large cities, identified short and long-term opportunities and threats to tourism development. They found that the impact of political instability on reducing tourist flows was most significant in the regions neighboring the aggressor country. These regions and cities were the objects of tourism for the inhabitants of the aggressor country. Their flow reduced sharply. However, tourism development continues in distant conflict regions and cities with a developed tourist background.

Iastremska, Kononova [20] investigated the short-term impact of the COVID-19 pandemic in 2019-2020 on international tourism worldwide, including separate regions of the world in particular. The main tendencies of change of directions and quantitative parameters of tourist flows at different pandemic stages are revealed. Activities to restore the tourism industry are proposed, forecasts of recovery dates are made. The main features and problems of the tourist sphere of Ukraine are revealed.

Summarizing the analysis results, we should note an in-depth study of the impact assessment of the COVID-19 pandemic on international tourism, forecasting scenarios for its future development, recommendations for overcoming the crisis, and recovery. However, the issues related to the analysis of the paradoxical increase in the influx of international tourists, mainly tourists from Arab countries to Ukraine during the COVID-19 pandemic, remained virtually unexplored. Their definition will allow us to assess the prospects for the restoration of international tourism in Ukraine and other countries in the future and purposefully manage these processes.

3 Purpose, aims, and research methods

The study aims to determine preconditions for increasing the tourists' influx from Arab countries to Ukraine in conditions of the COVID-19 pandemic, clarify their main preferences, and assess the prospects of formation and development of the international tourism market segment in Ukraine.

Consequently, the following research objectives were identified:

1) to determine preconditions for increasing the Arab tourists' influx to Ukraine during the COVID-19 pandemic, as well as the quantitative characteristics of the flow;

2) to determine the main characteristics and preferences of tourists from Arab countries visiting Ukraine;

3) to determine the prospects of formation and development of the international tourism market segment in Ukraine, which is focused on tourists from Arab countries during the pandemic COVID-19 and after its completion.

Backgrounds for the increasing tourists' inflow from Arab countries to Ukraine in 2021 and its quantitative characteristics were determined based on the analysis of latest information characterizing economic, social, environmental problems of the state and tourism regions, including analytical data of public authorities; data from specialized analytical agencies; open data of travel agencies, as well as tourist infrastructure facilities; media publications, etc. According to the analysis results using the synthesis method, the profiles of typical consumers of tourist services - residents of Arab countries who visited Ukraine were also compiled. The SWOT-analysis method was used to assess the prospects for developing the international tourism market segment in Ukraine focused on tourists from Arab countries.

4 Results of the study of the COVID-19 pandemic impact on the development of the international tourism market segment oriented to Arab tourists in Ukraine

4.1 Backgrounds for the increasing Arab tourists' inflow to Ukraine during the COVID-19 pandemic

According to travel agencies, the sharp increase in the flow of Arab tourists to Ukraine started with the Decree of the President of Ukraine of July 21, 2020, on establishing a visa-free regime for citizens of several Arab countries [21]. In January 2021, an agreement on air services between Ukraine and Saudi Arabia was signed. Currently, the number of flights is several dozen per week. The first flights took place to Kyiv. In July 2021, the Ukrainian "SkyUp" airline began charter flights from Saudi Arabia to Lviv and direct flights to Odesa. Flights to Kuwait (Jazeera Airways) were established, flights to Qatar were resumed.

A significant role played in Saudi Arabia in May 2021 repealing the ban on departure of its vaccinated citizens and those who had contracted the disease. At the same time (in May 2021), the 'Visit Ukraine' presentation (association of representatives of the tourist transformation of Ukraine) [22] with a group of Ukrainian tour operators took place at the Arabian Travel Market international exhibition in Dubai. Thus, tourists accustomed to travelling in Europe were presented with a new route. It was presented as an alternative to closed France, Switzerland, Spain, and other traditional Saudi Arabia tourists' vocation. Activities of state support to promote tourism in Ukraine for Arab countries' residents should also be mentioned.

An example is the meeting held on July 29, 2021, between the Minister of Culture and Information Policy of Ukraine and heads of diplomatic missions of Qatar, Kuwait, and Saudi Arabia in Ukraine. Its main topic was the expansion of cooperation in the field of tourism and culture [23]. Well-known bloggers from Saudi Arabia, who visited Ukraine and shared their positive impressions, also contributed to popularizing tourism in Ukraine [2].

Ukraine is an accessible and attractive country primarily for family vacations of Arab tourists. This is facilitated by:

- moderate and mild climate, which allows you to relax from the heat that is typical for this time of year in Arab countries, especially the Gulf countries;
- picturesque nature (forests, rivers, lakes), not typical for hot climate;
- large cities that tourists are accustomed to visiting in Western Europe, with exciting architecture, numerous parks, and attractions;
- interesting excursion programs and visits to natural, historical, ethnographic objects;
- significant potential for shopping, in particular, goods with a unique national flavour;
- affordable prices;
- delicious food, suitable for Muslim ethnic groups;
- friendly attitude of the population in general and the tourist infrastructure staff in particular.

These incentives, according to [24], caused an increase in the number of tourists from Saudi Arabia, UAE, Kuwait, Oman, Qatar, and Bahrain more than eight times in three months of 2021 (up to 35 thousand people) compared to 2019 (pre-crisis year). Moreover, in the first half of 2021, compared to the second half of 2020, the number of tourists from Saudi Arabia increased 40 times [25]. Tourists from Saudi Arabia, UAE, Oman are leaders of Ukraine among citizens of Arab countries (more than 85% of the total number of tourists). At the same time, one vacationer from the Persian Gulf countries, according to 'Visit Ukraine' [22, 24], spends an average of USD 2.5 thousand a week.

It should be noted that this is only a tiny proportion of Arab tourists who visited Europe in the pre-crisis period. According to the World Bank [26], in the pre-crisis years, the number of visits to Europe by tourists from only Saudi Arabia, respectively, was: 2015 - 21834000; 2016 - 20887000; 2017 - 18607000; 2018 - 17570000; 2019 - 20292000. That means that the tourist potential of the "Arab" segment of the international tourism market in Ukraine is very significant. Of course, not all Arab tourists will prefer Ukraine after repealing restrictions of European countries where they are used to rest. However, a certain proportion of them, especially those who have visited Ukraine, may remain loyal and influence others. In this context, it is essential to identify the prominent locations in which the bulk of tourists are concentrated and establish the reasons for this. This should be further taken into account when planning and stimulating the development of the segment of the international tourism market in Ukraine, focused on Arab tourists.

Geographically, most Arab tourists are concentrated in Kyiv, Lviv, and the Carpathians (Bukovel), Odesa. The analysis shows that these locations include almost all the tourists as mentioned above' interests; in addition, there is a large mass of five-star hotels in great demand. From these positions, other regions of Ukraine tendering similar services and conditions are advised to consider in the future. These are Kharkiv and Kharkiv region, Dnipro and Dnipropetrovsk region, Zaporizhia and Zaporizhia region. etc., which have 4–5-star hotels and facilities that may be attractive to Arab tourists.

4.2 Main characteristics and interests of Arab countries tourists who visited Ukraine in 2021

The generalization of the results of the analysis of information sources, mainly [2, 22, 3, 27, 28, 29, 30, 31, 32], etc., made it possible to consolidate the profile of Arab tourists who visited Ukraine in 2021. The leading indicators that are highlighted and taken into account in determining the profile, as well as their characteristics, are given in Table 1.

Table 1. Generalized profile of Arab tourists in Ukraine

Indicator	Characteristics
Type of tourism	Primarily family vocation: married couples; families involving children and often grandparents. Groups of friends meet
Income level	Average and above average
Duration of vocation	Mostly 7-14 days
Specific features of tourists	Religious people, whose behavior is determined by the requirements of the Muslim religion. In general, not conflicting
Experience in the level of tourist service	The vast majority have previously visited European countries and are well versed in the quality of tourist and related services
Accommodation requirements	Mostly five- and four-star hotels
Language of communication	Most speak English, but Arabic is also required.
Requirements for the service level	High: for the residence, food, transport, service level, taking into account the norms of their religion, etc.
Specific requirements for tourists	Large hotel rooms with several bedrooms; availability of certain types of equipment in bathrooms; large cars for family travel, etc.
Main interests	Entertainment (parks, attractions); natural landscapes, natural phenomena (including rain); shopping (interested in products of local flavour); excursions (monuments of nature, history, architecture, ethnography, etc.); gastronomic preferences (including local cuisine); travel to the regions of Ukraine, etc.

As it follows from Table 1, Arab tourists who used to spend summer vacations in western Europe went to Ukraine in 2021. They are well versed in peculiarities of recreation in Europe, at the level of service and prices. Naturally, they require a similar service in Ukraine. Or a fair price/quality ratio for travel and related services. In general, they perceive the prices of tourist services in Ukraine as low. A detailed analysis of the interests of Arab tourists shows that most of them are attracted by nature, shopping with local flavour, gastronomy. Among the shortcomings, they note a relatively low level of service, lack of inscriptions in Arabic and lack of Arabic-speaking staff, problems with food for Muslims (halal), bad roads etc. However, positive impressions from Ukraine prevail. They need to be consolidated so that this year's boom, caused by several favourable factors, is the beginning of long-term trends in the Ukrainian market of international tourist services.

The information listed in Table 1 can be used to develop a set of activities to improve tourism services and their rendering systems, which are focused on Arab countries' consumers. It can also be used in assessing the prospects for attraction of Arab tourists to other regions of Ukraine, mentioned in the previous section (Zaporizhia, Dnipropetrovsk, Kharkiv regions, etc.). In this case, the Table 1 characteristics of the generalized Arab tourists' profile should be compared to characteristics of the components of the tourist potential of promising regions (mentioned above and others), in particular [33]: recreational, resort, historical, ethnographic, climatic, religious, ethnic, etc.

4.3 Prospects for the development of the international tourism market segment focused on Arab tourists in Ukraine

Prospects were evaluated by the SWOT-analysis method. Table 2 outlines Ukraine's tourism potential's strengths and weaknesses identified according to the consumer demand of Arab tourists (see Table 1). It should be noted that their definition was made for regions visited in 2021 by the majority of Arab tourists.

Table 2. Strengths and weaknesses of the tourist potential of the regions of Ukraine in the context of interests of Arab countries tourists

Strengths	Weaknesses
Picturesque nature, attractive for tourists	Insufficient promotion of tourist services in cities and regions with high conditions for Arab tourists: Dnipro, Zaporizhia, Kharkiv, etc.
Mild climate (coolness and humidity), natural phenomena that are not typical for Arab countries	
Cities with numerous architectural and historic sites, parks	
Four- and five-star hotels with friendly and communicable staff	The low number of four- and five-star hotels, as well as suites in existing hotels, taking into account peculiarities of Arab tourists (see Table 1)
Visa-free entry to Ukraine	The low number of suitable autos for Arab tourists travelling with large families
Direct aircraft connections with Arab countries, as well as between central tourist regions of Ukraine	Poor roads condition
Low prices	Attempts at price discrimination (overpricing)
The friendly attitude of the population of Ukraine	Insufficient English and Arabic speaking staff at hotels, restaurants, tourist facilities.
Extensive opportunities for shopping (world brands, as well as goods with a national flavour)	Lack of Arabic inscriptions in hotels, restaurants, tourist sites.
Delicious food, including Ukrainian national cuisine, taking into account the peculiarities of Muslim tourists' demand	Few food establishments with a halal certificate
Many natural, historical, architectural, ethnographic, etc. objects (excursion or independent visit)	Few Arabic speaking guides
Numerous objects for entertainment	

As follows from Table 2, Ukraine has significant potential for developing the international tourism market segment focused on Arab tourists. Reducing weaknesses is mostly the incompetence of tourism service providers and business entities involved in this process (economic contractors of tourism service providers). However, the correction of some significant shortcomings is a responsibility of state and regional authorities, in particular: roads and highways, regulation of the hotel certification system, the food certification system, and so on.

However, the prospects for international tourism development in Ukraine are primarily determined by external factors. At present, they are favourable, but both short-term and long-term perspectives should be analyzed. Table 3 presents the results of the analysis of external opportunities and threats in the context of developing the international tourism segment in Ukraine focused on Arab countries tourists.

As it follows from Table 3, favourable opportunities are primarily caused by the current state of constituent subsystems of the external macro-environment. The threats are related mainly to possible adverse changes, studied from the standpoint of the analyzed segment of the international tourism market in Ukraine.

Table 3. Opportunities and threats for the development of international tourism in Ukraine focused on Arab countries tourists.

Opportunities	Threats
Visa entry to European countries	Repealing or reduction of COVID restrictions in Europe
European COVID-restriction for foreigners.	
Repealing the ban on departure of its vaccinated citizens and those who had contracted the disease in Saudi Arabia	Intensifying COVID-restrictions in Ukraine is possible
Positive impressions of Arab tourists who visited Ukraine in 2021 combined with the habit of vacationing in Europe	The threat of political and/or economic crisis in Ukraine

Thus, the combination of external and internal factors indicates significant prospects for forming and developing the international tourism market segment in Ukraine focused on Arab countries tourists.

5 Discussion

Determining the external and internal backgrounds for the 2021 growth of Arab tourists' influx to Ukraine allowed us to outline the range of competitive advantages that are significant for this group of consumers of tourism services in a COVID-19 pandemic. It is shown that competitive advantages are formed on: Ukraine's government and tour operators' successful use of the situation in the field of international tourist services, concerning Arab tourists, which allowed Ukraine to position itself as an alternative to closed European countries; strong tourist potential in several regions of Ukraine, suitable for Arab tourists and conditions like in European countries - internal prerequisites. These preconditions marked the beginning of the formation and development of a promising segment of the international tourism market in Ukraine, focused on Arab tourists. This is evidenced by the quantitative characteristics of the flows of these groups of tourists. The possibility of extending or forming a segment of the international tourism market to other regions of Ukraine is shown. Thus, the possibility of formation and development of new segments of the international tourist services market in the conditions of the COVID-19 pandemic is substantiated, which is connected with the change of directions of tourist flows.

The characteristics of the generalized profile of Arab tourists who visited Ukraine in 2021 form the information background for developing an activity system aimed at improving tourism and related services and the system of their rendering. They can also be used to assess the prospects of attracting Arab tourists to other regions of Ukraine. Thus, the profile data of tourists from Arab countries provide relevant information for effective management decisions in the tourism sector both at the level of tourism service providers and regional and state authorities. A particular disadvantage is that the primary sources of secondary information that summarize the data of individual providers of tourism and related services were used in compiling the profile. However, a specified profile gives a complete picture of tourists' preferences, allows to predict their behaviour in consumption and communication, which provides an opportunity to prepare tourist products, provide appropriate conditions for tourism services, train staff interacting directly with tourists, purposefully form customer loyalty. This will ensure Arab tourists' positive impression from a vocation in Ukraine, increase chances of their re-visits, and share positive impressions among potential consumers of tourist services.

The assessment of the possibilities of formation and development of the international tourism market segment in Ukraine focused on Arab tourists, performed by the SWOT-analysis method, testifies significant prospects. The COVID-19 pandemic provided a chance that was used successfully. According to the SWOT analysis results, Ukraine has a strong tourist potential and favourable external conditions (opportunities) for its implementation. In order to maintain the existing positive trends in the development of the international tourist market segment focused on Arab tourists in Ukraine, it is necessary to reduce and amend shortcomings identified by the analysis, which negatively affect the domestic opportunities of the country's tourism sector. However, it is necessary to anticipate the possible consequences of implementing identified external threats and develop activities to avoid or minimize them.

6 Conclusions and prospects for further research

The set of external and internal preconditions is determined and their impact on the 2021 growth of the Arab tourist's flow to Ukraine in conditions of COVID-19 pandemic restrictions introduced in most countries. Quantitative characteristics of this growth are outlined, indicating the beginning of the formation and development of the international tourism market segment in Ukraine, focused on Arab tourists. They also testify to the significant economic prospects of Ukraine's tourism industry. The main characteristics and preferences of tourists from Arab countries who visited Ukraine are determined. A generalized profile of Arab tourists is compiled. It shows that tourists who used to spend vacations in other European countries treat Ukraine as an alternative to traditional summer trips. An assessment of the prospects for formation and development of the international tourism market segment in Ukraine, focused on Arab tourists, in the context of the COVID-19 pandemic and after its completion was performed.

The obtained results can be used for management decisions at the state and regional levels of tourism development management and the level of individual tourism service providers.

Further research should be aimed at accumulating and analyzing information for 2022 and subsequent years, which characterize trends of formation and development of the international tourism market, focused on Arab countries tourists. It is also necessary to analyze the impacts on the tourism sector caused by the COVID-19 pandemic. This will clarify existing trends and outline the chances for success in the formation and development of this segment of the international tourism market in the future, including the post-pandemic period.

References

1. *The COVID-19 pandemic and its consequences in the field of tourism in Ukraine. Update to the document "Roadmap for competitive development of tourism in Ukraine".* Hotel & Destination Consulting. (2020) URL: http://www.ntoukraine.org/assets/files/EBRD-COVID19-Report-UKR.pdf.

2. B. Miroshnychenko, M. Tsymbaliuk. *Arab tourists have flooded Ukrainian resorts: what attracts them to Ukraine and how to use it.* Ekonomichna Pravda, 21 Jule (2021). URL: https://www.epravda.com.ua/publications/2021/07/22/676156/.
3. *Join UP!.* Official site. URL: https://joinup.ua/.
4. J. Abbas, R. Mubeen, P.T. Iorember, S. Raza, G. Mamirkulova. Current Research in Behavioral Sciences. 2 (2021). https://doi.org/10.1016/j.crbeha.2021.100033.
5. N.G. Uğur, A. Akbıyık. Tourism Management Perspectives. 36 (2021) https://doi.org/10.1016/j.tmp.2020.100744.
6. J. Saha, S. Haldar, S. Bhattacharya, P. Suman. Tourism in retrospect of COVID-19 on global perspective using analytical hierarchy process. Spatial Information Research. (2021) https://doi.org/10.1007/s41324-021-00407-4.
7. M. Aronica, P. Pizzuto, C. Sciortino. The World Economy. 00. 1-15 (2021) https://doi.org/10.1111/twec.13157.
8. E. Neumayer, T. Plümper, M. Shaikh. *The logics of COVID-19 travel restrictions between European countries.* Social Science Quarterly. (2021) https://doi.org/10.1111/ssqu.13016.
9. E.M. Gegung. Jurnal Kepariwisataan Indonesia: Jurnal Penelitian Dan Pengembangan Kepariwisataan Indonesia, **15(1)**, 9-15 (2021) https://doi.org/10.47608/jki.v15i12021.9-15.
10. C.-M. Cretu, A.-G. Turtureanu, C.-G. Sirbu, F. Chitu, E. S. Marinescu, L.-G. Talaghir, D.M. Robu. Sustainability. **13**, 8405 (2021) https://doi.org/10.3390/su13158405.
11. Y. Kitamura, S. Karkour, Y. Ichisugi, N. Itsubo. Sustainability, **12 (24)**, 10302 (2020). https://doi.org/10.3390/su122410302.
12. M-C. Tsai. PLoS ONE. 16(3), e0248319 (2021). https://doi.org/10.1371/journal.pone.0248319.
13. D. Binh, N. Ninh, D. Clare, D.B. Huu, N.H. Thi. Tourism and Hospitality Research, 0(0), 1–13 (2021). https://doi.org/10.1177/1467358421993902.
14. M.K. Rahman, M.A.I. Gazi, M.A. Bhuiyan, M.A. Rahaman. PLoS ONE, **16(9)**, e0256486 (2021). https://doi.org/10.1371/journal.pone.0256486.
15. R. Hamid, N.H.M. Hashim, S.A.M. Shukur, N.H. Marmaya. International Journal of Academic Research in Business and Social Sciences, **11(16)**, 27–41 (2021). http://dx.doi.org/10.6007/IJARBSS/v11-i16/11213.
16. O. Prokopenko, V. Omelyanenko. TEM Journal, **9(3)**, pp. 1068-1075 (2021) http://doi.org/10.18421/TEM93-31.
17. J. Cheng, Z. Zetina. Open Journal of Social Sciences, **9**, 326-334 (2021). https://doi.org/10.4236/jss.2021.97023.
18. V. Pashkus, N. Pashkus, A. Asadulaev. SHS Web of Conf., **92**, 01041 (2021). https://doi.org/10.1051/shsconf/20219201041.
19. N. Tomczewska-Popowycz, Ł. Sustainability, **13**, 4126 (2021). https://doi.org/10.3390/su13084126.
20. O. Iastremska, O. Kononova. Three Seas Economic Journal, **2(2)**, 54-61 (2021) https://doi.org/10.30525/2661-5150/2021-2-9.
21. Decree of the President of Ukraine №289 / 2020. On the establishment of a visa-free regime for citizens of Australia, New Zealand, the Kingdom of Bahrain, the State of Kuwait, the Sultanate of Oman, the Kingdom of Saudi Arabia (2020). URL: https://www.president.gov.ua/documents/2892020-34449.

22. Visit Ukraine. Official site (2021). URL: https://ukraine.ua/visit/.
23. ICIP intensifies cooperation with Arab countries in the field of tourism and culture. Official site of the Ministry of Culture and Information Policy of Ukraine (2021). URL: https://mkip.gov.ua/news/5808.html.
24. B. Miroshnychenko. *Ukraine received 35,000 tourists from the Persian Gulf in three months.* Ekonomichna Pravda, 2021, 17 August (2021). URL: https://www.epravda.com.ua/news/2021/08/17/676959/.
25. State Agency for Tourism Development in Ukraine (2021). URL: https://www.facebook.com/DARTUkraine/photos/a.114995336794002/366976171595916.
26. The World Bank. International tourism, number of arrivals – Saudi Arabia (2021). URL: https://data.worldbank.org/indicator/ST.INT.ARVL?locations=SA.
27. D. Kurishko, G. Erman. *How and why Arab tourists fell in love with Ukraine*. BBC News Україна, 2021, 7 August (2021). URL: https://www.bbc.com/ukrainian/features-58120247.
28. G. Marushchenko. *Arab tourists in Ukraine: is the Ukrainian coast ready for guests from the Persian Gulf.* O, More, 2021, September 10 (2021). URL: https://omore.city/articles/164482/arabski-turisti-v-ukraini-tradicii-vidpochinku-na-uzberezhzhi.
29. G. Tereshchuk. *Lviv became famous in the richest country in the world. But for how long?* Radio Svoboda (2021). URL: https://www.radiosvoboda.org/a/karantyn-turyzm-saudivska-araviya/31393411.html.
30. D. Kolomiets. *$40 million to the state treasury. How the Arabs raised the tourism business of Ukraine.* Suspilne novyny, 2021, August 28 (2021). URL: https://suspilne.media/159200-40-miljoniv-dolariv-u-derzavnu-kaznu-ak-arabi-pidnali-turisticnij-biznes-ukraini/.
31. *Up to 800 euros per day: tourists from Arab countries in Ukraine spend more Europeans* - the media. Focus, 2021, September 18 (2021). URL: https://focus.ua/uk/ukraine/493081-do-800-evro-v-sutki-turisty-iz-arabskih-stran-v-ukraine-tratyat-bolshe-evropeycev-smi.
32. *Arab tourists in Ukraine: where and how much money do they bring?* Infographics. Baker Tilly Ukraine, 2021, September 17 (2021). URL: https://bakertilly.ua/news/id50120.
33. S.M. Illiashenko, Y.S. Shypulina, N.S. Illiashenko. Marketing and digital technologies, **5(2)**, 14-26 (2021) https://doi.org/10.15276/mdt.5.2.2021.2

The Strategy Focuses of Recreation and Tourism Activities of Ichnia National Nature Park (Ukraine) Design Employing Canvas Business-model and SWOT-analysis

Viktoriia Kiptenko[1*], and *Tetiana* Mykhailenko[2]

[1,2]Department of Regional Studies and Tourism, the Faculty of Geography, National University of Kyiv named after Taras Shevchenko, 64/13 Volodymyrska str., Kyiv, 01601, Ukraine

Abstract. National parks across the world contribute to the preservation of biodiversity, playing simultaneously an outstanding role in sustainable activities, recreation and tourism in particular. Experience with COVID-19 shows that resilience is fundamental to the sustainability of protected area visitation, which was affected in different ways by pandemics. The last decade challenges for the national nature parks of Ukraine, positive impetuses of several successful projects, enhancement of the community role in the environmental issues caused the search for a framing of the strategy of recreation and tourism activity of the national nature park located in Ichnia, Chernihiv oblast, Ukraine. By employing the Canvas business model and SWOT analysis the client-oriented shift in the above park strategy focuses ground prospects for facts-based decision-making and planning. The focuses advised include restructuring of the organizational patterns of the entity; modernisation of information and communication activities; profiling of demand and diversification of values for travellers of different motivation; networking with stakeholders. Keywords: Recreation, Tourism Activities, Ichnia National Nature Park, Canvas Business-model, SWOT-analysis.

Introduction

Worldwide national parks have furnished models for biodiversity preservation involving sustainable tourism development and provided a blueprint to follow. In particular, frames of destination stewardship arm national park keepers with means to manage nature conservation with the needs of local people and tourists alike [1].

COVID-19 affected protected areas tourism in different ways, including the crucial decline of visitation, just starting to recover; rather slow and not straightforward economic recovery; internalisation of safety considerations into tourist management; examples of online ways of sharing nature; emerging innovations in ways of tourist-dependent communities support; creating stronger alternative revenue streams in tourism destinations

* Corresponding author: kipavika@gmail.com

by product diversification and innovations; gradual adaptation of travel and tourism market to the new circumstances [2].

For decades, the national nature parks (NNP) of Ukraine faced different problems inhibiting their potential for conventional environmental activities and development of nature recreation and tourism [3,4]. Protected areas like Gorgany reserve, Carpathian Biosphere reserve enjoyed the support of international aid and grants, which are still not enough for over 50 NNPs in the country. They mostly suffer from a lack of public funding, inability to arrange for revenues streams by themselves and improper - often outdated soviet style of - management.

The Ichnia NNP's, (situated in Chernihiv oblast, management) - in an attempt to innovate outdated methods of operation - enquired about the possibility to recast its strategy, searching, in particular, for ways to enrich its revenue streams by new vision and approach to recreation and tourism activities relevant to the environmental mission of this protected area.

Thus, we aimed our research on designing the strategy focuses for Ichnia NNP recreation and tourism activities. We speculated the relevance of employing the CANVAS business model and SWOT analysis to provide for a coherent vision of the state-of-art situation and substantiation of the client-oriented focuses for further decision-making and planning.

This study used both secondary and primary data. The secondary sources included legislative documents; reports and informational bulletins of Ichnia NNP; strategies of development for Chernihiv oblast, Ichnia and Parafievka united territorial communities (where the entity is located); relevant academic papers, and open-access information. We collected primary data through observations, field trips, semi-structured interviews with Ichnia NNP's management, as well as a micro-questionnaire of potential visitors/tourists.

Based on content analysis of available secondary sources and primary data exploration we designed the Canvas template to grasp the state-of-art of recreation and tourism activities of the park. We further applied the SWOT-analysis in order to articulate the strategic focuses for recreation and tourism activities decision-making and planning by Ichnia NNP's management.

Being aware that the results of this single case study has limited potential for generalisation, we believe that this example overcomes a lack of previous research discussing the business model of national nature parks and serve similar situations, in which the visualisation of the state-of-arts will help to frame strategic focuses, decision-making and planning of development to be supported by further monitoring and analysis of NPPs activities.

1 Results and discussion

1.1 Employment of Canvas business model and SWOT analysis fit the needs of Ichnia NPP's

The majority of NNPs of Ukraine, Ichnia NPP in particular, faces various problems on its way to improve the quality and the efficiency of operation. The challenges include illegal land spots invasions, poaching and illegal logging, irregular mosaic land use added by several authorities' initiatives on infrastructure that destroys protected areas, improper management aggravated by poor staff wages, unregulated activities and devastation of national nature parks in occupied Eastern parts of the country and the annexed Crimea, etc. [3,4].

The key possible solutions foresee the improvement of legal, organizational and economic activities in line with sustainable development grounds to increase the competitiveness of the tourism product and the quality of services offered by the national parks, which are basically publicly funded in Ukraine [5, 6, 7]. In the above sense, the prospects for the development of environmental-friendly tourism in NPPs has to follow the principles of ecological awareness, focus on nature, preservation of the pristine environment and the local socio-cultural heritage; economic efficiency and sustainable development of the areas where the tours take place [7]. That is why the provision of these principles and the implementation of various strategies could be based on experience, which demonstrates the high competitiveness of national parks having revenue streams from recreation and tourism activities [8, 9].

Considering that the business model has an important role in an entity's strategy and explains how their activities operate jointly to implement their strategy, we followed the notion that the business model was studied as a potential tool at various levels of the product, for business units, companies, and industries involved in nature tourism [10]. In addition, broadly used in business and management the business-model concept interpreted through sustainability concerns can be used to achieve the triple-bottom result of short- and long-term perspectives essential for the NPP: economic, social and environmental [10,11].

The Canvas business model is uncommon for nature recreation and tourism research. Most often, this model suggests new opportunities for tourism business management, ways of strengthening competitive positions and core values for other types of tourism: diving [12], sports [13,14], beach [15] and sailing [16], creative [17], as well as to create a tourist city branding [18].

We consider the opinion that Canvas is only a tool that enables firms to articulate their business model [19] advantages Ichnia NNP's administration which lacks appropriate experience. The visualising potential of Canvas' nine blocks [20] - consumer segments, key values, customer relationships, distribution/channels, key resources, key activities, key partnerships, cost structures, and revenue stream - fits the enquiry of Ichnia NNP to client-oriented shift in their recreation and tourism activities to create revenue streams. It is the key values that underlie this model, form a priority in this model and factor customers attraction [19].

SWOT-analysis allows to overcome the critics against static reflections of the situation by Canvas business model and helps to obtain several alternative strategies [21]. By identifying the strengths and weaknesses of the NNP, assessing strategy through opportunities and taking into account threats, we synthesised the relationship between "internal" (strengths and weaknesses) and "external" (opportunities and threats) factors. The formation of a matrix of opportunities and threats with their subsequent ranking - based on the division into three groups according to the probability and degree of influence on the organization of recreational and tourist activities of the NNP allows to characterize the extent to which opportunities promotes strengths, and whether threats do not level a particular strength and the opportunities provided for it. The outcomes shape comparative advantages, challenges and risks that determine the basis for strategic choice - the formulation of strategic objectives for the development of the NNP in the long run.

1.2 The combined Canvas-SWOT design for Ichnia NPP's recreation and tourism activities shapes the state-of-art and beyond.

Ichnia NPP was established in 2004 and has a total area of 9665,8 ha; it belongs to the Emerald network.

According to its natural potential, the territory of the park has no alternative in the northern part of the Dnipper left bank forest-steppe. Mostly flat-wavy relief, temperate continental climate with mild winters and warm summers, the Udai and Ichenka rivers, which form a kind of wetlands, numerous ponds and lakes, the richness and value of flora and fauna, preserved in its natural state, contribute to the comfortable conditions for recreation and tourism. They provide for environmental and nature-oriented tourism involving walking, excursions (hiking, biking, skiing), boating, mushroom and berry picking, fishing, hunting, etc.

At the same time, climate change in recent years caused increasing arid conditions in the summer. Respectively, hydrological resources lose their recreational attractiveness. Thus, organization of recreational and tourism activities needs consideration and flexibility.

The zone of regulated recreation constitutes 50.5% of the NNP's territory and the zone of stationary recreation amounts for 1% of it. More than 50% of the territory is in permanent use without seizure from land users (Zhadkivske, Kamyanske and Ivanytske forestry). Administratively, the territory of the NNP belongs to Ichnia and Parafiivka united territorial communities of Prylutsky district, Chernihiv region.

The territory of the park and its surroundings have a rich nature and ancient history, which raises the potential for the organization of recreational and tourism activities. In general, the Ichnia area has preserved numerous monuments of history and culture, which improve the potential of the Ichnia NNP' locations as sites of cultural, historical, ecological, adventure and other tourism routes.

These are mounds of the Bronze Age and the Scythian period, Chernyakhiv culture, ancient Slavic settlements, rich history of the Cossack era, architectural and park ensembles, architectural monuments, art centers and museums, including those associated with prominent artists. The historical and cultural heritage reflects the fact that the city of Ichnia in the past was a prominent center of folk arts and crafts: pottery, weaving, and embroidery. Painted ceramics of Ichnia potters in the 18th century reached a high artistic level. At the end of the 19th century Ichnia produced the best pottery in Ukraine, which was in demand far beyond the region. Ichnia tiles, unique works of folk decorative art, were especially famous. Some of them can be found in the Ichnia Museum of Local Lore.

NPP's staff interview and reports suggest that in 2020 they had just 384 visitors who participated mainly in environmental educational activities despite several ecological trails and routes designed and about 22 places furnished for temporal recreation.

The combined Canvas and SWOT template reflects generalized results of interviews, observations and surveys (Table 1). It further serves to articulate the strategic focuses of client-oriented recreation and tourism activities of Ichnia NNP as follows:

Restructuring of NNP's organizational patterns by:

- establishing of visit-center to perform, among others, marketing activities;
- initiating entrepreneurial activities to develop environmental educational, recreation and tourism activities.

Modernization of information and communication activities by using modern digital tools:

- to develop and implement the communication and marketing plans;
- to modernise and integrate the official web-site with travel platforms and modern messengers (Instagram, Telegram, Viber, TikTok);
- to promote the new image values in the context of recreational and tourism activities (i.e. SafeTravel, Climate and Plastic initiatives of UNWTO, some already supported by the National Tourism Organisation of Ukraine);
- to develop production of souvenir and promotional products.

Profiling of demand by:

- exploring the profile of NNP's visitors;
- monitoring of innovations in the market of nature tourism services;
- monitoring of innovations and exchanging experience with NNPs of Ukraine;
- determining the demand of stakeholders/partners (communities, civil society organizations, rural tourism farmsteads, tour operators, etc.).

Table 1. Combined Canvas-SWOT design for Ichnia NPP's recreation and tourism activities

Strengths	Weaknesses	Opportunities	Threats
1	**2**	**3**	**4**
CANVAS element - Consumer segments			
Stable segment of consumers from educational institutions of different degrees	Narrow (age, geography) segment of consumers, lack of defined consumer profiles, a meager share of nature and environment-oriented tourists with various motivations for travel.	To design consumer profiles, structure segmentation, monitor.	General trends of depopulation, unfavorable socio-economic situation in the country and the region
CANVAS element - Key values			
Very basic experience of services provision	Low level of - values supply correlation with potential demand; integration of ecologic educational activities into recreational and tourism services; image elements exploitation; - regulation of anthropogenic load from spontaneous use of places for short-term recreation; - coordination of interests in the formation of values between land users; - coordination of interests in the values formation with stakeholders; - supply diversification (seasonal for some types of activities (cycling, foreseen only for students); - supply of values / services for people with disabilities.	To identify and develop diversified values (services/ products)	Low level of population awareness and motivation, inability to diversify values agreed with stakeholders due to conflicts of interest and due to underfunding/delays in the overhaul of the eco-center.
CANVAS element - Distribution/channels			
Some experience of internet and social media usage for communication, updating of information in FB (1010 followers); the primary information infrastructure on the territory	inefficient system of communication channels (limited access to printed materials), imperfect content of the web-site, one page in one social media, lack of tools for feedback, visualization and interactivity; imperfection of information infrastructure in the park; lack of channels for people with disabilities.	To diversify and modernise communication channels; to develop and systematically perform the communication plan.	A lack of adequate ICT coverage; inability to independently diversify and carry out communication activities; inadequacy of communication plans and actions; loss of competitive advantages of the park.

Table 1. (continued)

1	2	3	4
CANVAS element - Customer relationships			
Developed formats of direct relations with educational institutions and some local public organizations	lack of a visit center, marketing plan and "automated" communication with consumers; printed promotional materials have any appropriate marketing usage; insignificant interaction with stakeholders; any cross-marketing communications.	To design and implement marketing plan	institutional / legislative and funding constraints, conflict of interest with partners; erroneous marketing and promotional campaigns.
CANVAS element - Key resources			
The park's natural potential has no alternative in the northern part of the Left-Bank Forest-Steppe	scattering of territories across Ichnia district and mosaic patterns of the park and other land users spots; remoteness of places for rest and significant distance to the dendrological park "Trostyanets"; imperfection of information, transport, infrastructure and sanitary and hygienic provision; the historical and cultural attractions of the adjacent territory avoid including NNP's locations in the product; lack of labor force for recreation and tourism, marketing and promotion activities; lack of amenities for disabled people.	To modernize the potential of the park resources for the creation of values / services of modern types of recreation, including for people with disabilities; integration with the historical and cultural heritage and resources of the surrounding areas; digitization of resources.	further deterioration of the climate conditions, anthropogenic load, pollution, fire danger, lack of full-fledged ICT coverage; lack of funding.
CANVAS element - Key activities			
Operational eco-center, ecological-educational activities, supported by scientific research, rendering of additional services, maintenance of 22 places for short-term recreation.	A lack of a visit-center and systematic activities on marketing and promotion, study of demand, integration of eco-educational and recreation and tourism activities, participation in projects; systemic partner-network interactions.	To develop organizational patterns to ensure the visit center, marketing and promotional activities, diversification of opportunities for the use of resources and the creation of values corresponding to demand; interaction and use of cross-marketing tools in partnership with stakeholders.	Inability to modernize activities; lack or delay of funding for activities and innovations.

Table 1. (continued)

1	2	3	4
CANVAS element - Key partners			
Strong partnership with educational institutions and some public organizations of Ichnia area, some experience of interaction with tour-operators.	Insignificant number of strong partners limited to local level entities, weak relations with other nature reserves of Chernihiv region / Ukraine, almost no interaction with stakeholders (land users, communities, tour-operators, rural tourism farmsteads, etc.).	To network and establish collaboration, cooperation and long-term partnerships with stakeholders, focused on opportunities of cross-marketing and clustering, public-private partnership.	Conflicts of interest (competition) with stakeholders/partners, impossibility of concluding long-term agreements, fear to go beyond the local scale.
CANVAS element - Costs			
Fixed costs are covered by funding from the state budget	Irrational costs for obviously uncompetitive products or services; unstable variable costs; Tiny revenue streams limit additional expenses for arrangement of the territory, production of souvenirs, etc.	To generate revenue streams in order to cover (partially) maintenance and activities costs.	A lack of funding or underfunding of NNP.
CANVAS element - Revenue streams			
Legislatively provided ability to independently set prices for recreational services that the national park provides to tourists/	Not flexible pricing policy; limited income from the existing accommodation facilities for the creation of values/services; any sponsorship, grant and project activities experience.	To optimise pricing policy, revenue streams through diversification of services and the establishment of strong partnerships, involvement into project, grants and sponsorship.	inability to realize opportunities to generate revenue streams, low solvency of the population.

Diversification of values for a wide range of customer segments with different travelling motivation by:

- optimising the existing and designing new trails, parking lots, routes to provide environmental educational and nature recreation and tourism services;
- modernising existing and creating new infrastructure (including temporal and permanent accommodation) for sustainable recreational and tourism activities, ensuring the safety of visitors and protection of nature;
- creating locations within NNP's territory attractive for inclusion into tourism products across the neighboring territory.

Networking with stakeholders (communities, civil society entities, tour-operators, rural farmsteads, etc.) based on public-private partnership models and usage of cross-marketing tools, including:

- identification of common interests related to the values of NNP and development of a joint action plan for the implementation of development strategies and programs;

- participation in pilot project of regional tourism route "The Taste Routes of Chernihiv Region";
- promotion of establishment of rural tourism farmsteads in villages located within the territory of the park and neighboring areas;
- participation in grant programmes and projects.

2 Conclusion

Current challenges of national parks development in Ukraine require an appropriate recasting of their strategies to overcome outdated soviet-time management fails. Recreation and tourism activities have potential to facilitate revenue streams and can rely on employing Canvas business model combined with SWOT-analysis in design of strategy for their development. This research of Ichnia NPP presents a case study to overcome a lack of previous explorations of that kind and serve as a ground for further facts-based decision-making and planning.

References

1. How national parks around the world influenced sustainable tourism development. World Travel and Tourism Organization.. July (2016). URL: https://worldtraveltourismcouncil.medium.com/how-national-parks-around-the-world-influenced-sustainable-tourism-development-6e149cfc0688
2. A. Spenceley, S. McCool, D. Newsome, A. Báez, J. R. Barborak, Cl.-J. Blye, K. Bricker et al. Parks, 27 (Special issue) (2021) URL: https://parksjournal.com/wp-content/uploads/2021/03/Spenceley_et_al10.2305-IUCN.CH_.2021.PARKS-27-SIAS.en_.pdf
3. V. Mhlamian, Natsionalni parky Ukrainy: nishchivne ruynuvannia zapovidnykh zon (National parks of Ukraine: devastating destruction of protected areas), UNIAN. January (2016) URL:https://www.unian.ua/ecology/1250448-natsionalni-parki-ukrajini-nischivne-ruynuvannya-zapovidnih-zon.html (In Ukrainian)
4. N. Konishcheva, S.Tkachova, Innovations and technologies in the service sphere and food industry. **2**, 23-35 (2020) (In Ukrainian)
5. O.Milinchuk. Elektronnyi naukovyi zhurnal «Pryazovskyi ekonomichnyi visnyk», **5**, 67-71 (2017) (In Ukrainian)
6. Yu.Onoyko, Naukovyi visnyk Khersonskoho derzhavnoho universytetu. Seriia: Heohrafichni nauky, **9**, 154-157 (2018) (In Ukrainian)
7. N. Pankiv, Scientific Bulletin of UNFU, **29(5)**, 88-92 (2019) (In Ukrainian)
8. T. Bezsmertniuk, M. Melniichuk, Naukovi zapysky Ternopilskoho natsionalnoho pedahohichnoho universytetu imeni Volodymyra Hnatiuka. Ser. Heohrafiia. Ternopil: Taip, **1(48)**, 77–83 (2020) (In Ukrainian)
9. S. Korotun, M. Yakovyshyna, Visnyk NUVHP, **2(74)**, 163-170 (2016) (In Ukrainian)
10. S. Sahebalzamani, G. Bertella, Sustainability, **10**, 1–15 (2018)
11. B. Wirtz. Business Model Management: Design, Instruments, Success Factors. Gabler: Wiesbaden, Germany (2011)
12. R. Wingit, E. Maulina, M. Rizal, M. Purnom, International Journal of Management and Administrative Sciences (IJMAS) (ISSN: 2225-7225) **4(12)**, 45-64 (2017)

13. M. Perić, N. Wise, D. Dragičević, Sport, Business and Management, **7(1)**, 58-76 (2017)
14. M. Peric, N. Wise, Local Economy. **30**, 1000-1016 (2015)
15. Mirza, International Journal of Innovative Science and Research Technology, **4**, 703-710 (2019)
16. R. Strulak-Wójcikiewicz, N. Wagner, A. Łapko, E. Hącia, Procedia Computer Science, **76**, 1643-1651 (2020)
17. G. Richards, Journal of Hospitality and Tourism, **19(1)**, 1-13 (2021)
18. S. Debarliev, S. Mitrovska, European Scientific Journal, **12(35)**, 82-104 (2016)
19. R.Ambrož, D.G.Omerzel, Academica Turistica, **10(2)**, 175-183 (2017)
20. A. Osterwalder. *Business Model Generation*. John Wiley&Sons, Inc, Hoboken, Ney Jersey, (2010)
21. F. R. David. *Strategic Management concept and cases* (13th ed). United Stated of America: Pearson Education, INc., Publishing as prentice Hall (2011)

Customer Engagement with Universities' Social Media after COVID-19: EaP Countries Case

Nina Chala[1], *Kateryna* Pichyk[1], *Olga* Voropai[1], and *Olena* Bilovodska[2*]

[1]National University of Kyiv Mohyla Academy, Marketing and Business Management Department, 04070, 2 Skovorody Street, Kyiv, Ukraine
[2]Taras Shevchenko National University of Kyiv, Department of International Economics and Marketing, 01033, 60, Volodymyrska Street, Kyiv, Ukraine

Abstract. The current research aims at analyzing the social media activity of the universities in EaP countries. During the COVID-19 pandemic, universities had to transfer their education processes and enrollment campaigns online. University management had to introduce new tools for communication and interaction to ensure the efficacy of cooperation between the university community, students, teaching staff, and other stakeholders. In this research, the authors analyze which factors affected the dynamics and directions of customer engagement with Universities' social media and how. The authors propose a framework to reveal the role of countries' digital environment, educational environment, and university's SM activity in increasing ER of university stakeholders during the COVID-19 pandemic on Facebook and Instagram. The authors outline managerial proposals for universities, including analytics-based SM content creation and planning, communicating socially essential issues.
Keywords: universities' social media, COVID-19, Facebook, Instagram.

1 Introduction

Higher education institutions (HEI) widely use SM to communicate with their clients, increase brand loyalty, co-create value with students, academic staff, and other stakeholders. However, depending on the university's goals and objectives, the marketing role of SM use differs and is often underestimated.

Several studies investigate how universities use SM for public relations and university brand-building and include a qualitative, sentimental, and quantitative content analysis of the posts and corresponding users' Engagement rate (ER). HEI researchers and managers are interested in the role of customer engagement with Facebook and other SM pages towards prospective student enrolment. Several studies are devoted to SM usage by separate university departments. Ihejirika K., Goulding A., Calvert Ph. suggest applying a strategic approach to implementing SM into the academic libraries' marketing practice [1]. Roopchund R., Ramesh V., Jaunky V. C. indicate the effectiveness of introducing SM into the education process by faculty staff to enhance a higher level of students' communication and engagement [2]. All these activities become a part of a more strategic value co-creation

* Corresponding author: o.bilovodska@gmail.com

process. Value co-creation plays an essential role in attracting active participants to the university community. It helps to develop innovations and develop new services and products. The paper by Al-Kumain N. H., Alhazmi A., Rmayah T., Shabbir M. S., Gazem N. explores the use of online platforms for value creation by Malaysian universities, including crowdfunding platforms for attracting funding for innovative development [3]. Within the framework of the current research, we consider the processes by which universities create value for their clients on social networks, which were covered in the papers by Voropai O., Pichyk K., Chala N. [4] and Horodovenko V., Bilovodska O., Vatras V., Kanzafarova I., Melnyk R. [5]. However, we shift the focus of this paper to the COVID-19 period, which has forced universities to become more active in SM.

The impact of the COVID-19 pandemic on universities and their interaction with students is only in the spotlight of researchers. This can be exploited by the mass transition of universities to distance learning, which took place a year ago. This transition is large-scale and unprecedented as it has never been before. In addition, we are currently unable to envisage and report all the consequences of such a situation. For example, a study conducted by the Office for National Statistics found that students were more responsive to quarantine restrictions than ordinary citizens [6]. More than 29% of students surveyed reported a negative experience of distance learning, and more than 53% of students reported a deterioration in well-being and mental health due to distance learning. The numbers highlight the need for universities to interact with students.

Research on the impact of COVID-19 on universities focuses mainly on the effectiveness of distance learning and future models of blended and hybrid learning [7, 8]. Fenwick M., McCahery J. A., Vermeulen E. concentrate on accelerating the digital transformation and IT support of higher education [9]. Gezmen B., Eken I. analyze individual adaptation to online learning, time costs, workload, and the overall level of student satisfaction with new learning forms and tools [10].

Scarce research studies on HEI social media usage in post-COVID times consider communication in social networks in the context of the efficiency of the educational process and knowledge delivery via TikTok [11] and YouTube [12, 13]. The interaction between universities and their community, stakeholders, enrollees in the conditions of strict lockdown has not been studied. Therefore, the presented work is unique.

2 Materials and methods

We use our research to investigate how the COVID-19 pandemic influenced the established order of things in the interaction between universities and their stakeholders through social media. Fig. 1 depicts our conceptual framework, which views customer engagement as an outcome of such interaction. The COVID-19 pandemic and the corresponding local and global policy changes that equally severely affected all the countries in the region highly disrupted the whole process and its results. Nevertheless, we suppose that the level of reaction and changes differs in different countries. We consider the main factor that may have influenced such differences is institutions' activity in social media.

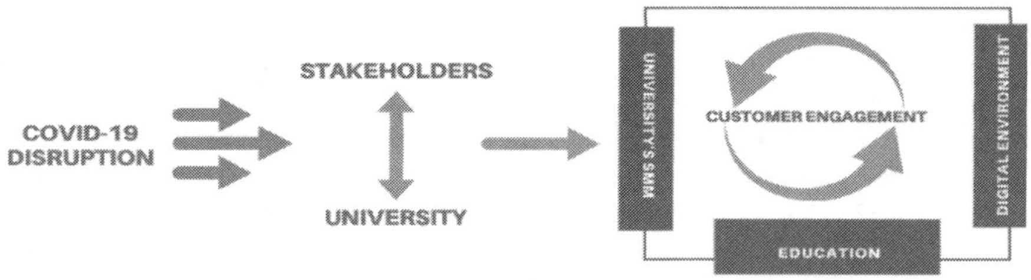

Fig. 1. Conceptual framework

In this study, we addressed the main research question:

Are social media activities of the universities in post-pandemic in line with what their stakeholders expect and are willing to engage with?

We selected the five most influential universities in each Eastern Partnership country based on the academic impact study*. An academic impact rating InfluenceRanking is created based on data analysis from open sources in different languages regarding citations, references, achievements of the university community [14]. Thus, this rating considers scientific achievements, publishing activity of university representatives, and recollections in mass media, social networks, etc. Machine learning algorithms are used for data analysis, which makes it possible to update the data quarterly.

The research included Facebook and Instagram as the most popular social media platforms in EaP countries, according to Datareportal [15].

We investigated two periods - March-September 2019 (pre-COVID-19) and March-September 2020 (COVID-19) - to compare the interaction of universities through social networks before and during the coronavirus pandemic. We considered only the official pages of universities on social networks. We excluded information and interaction with the audience of individual pages of teachers, administration, departments that do not have "links" to the official from our analysis.

We collected an array of data from the official pages of universities on social networks: the number of subscribers, posts, likes, shares, and comments*. We also extracted data on the level of SM engagement using Popsters software [16].

To track down the level of user engagement after COVID-19 in social networks, in this section, we will use the following metrics for Facebook and Instagram [16]:

- Number of fans
- Fans/students ratio
- Engagement rate per post
- Engagement rate per day
- Number of comments
- Number of likes
- Love rate
- Talk rate

Additional metrics were introduced to cluster universities and to check hypotheses on a country level:

- University posting rank. Each university is assigned points depending on its SM activity dynamics in Facebook and Instagram.

0 points - no activity.

-1 point - the page terminated.

1 point - negative posting dynamics.

2 points - positive posting dynamics.

- Country posting rank is a sum of average posting ranks for all five universities in a given country.
- University ER rank. Each university is assigned points depending on its users' activity dynamics on Facebook and Instagram.

0 - know university SM page or page terminated.

1 - negative ER dynamics.

2 - positive ER dynamics.

- Country ER rank is a sum of average ER ranks for all five universities in a given country.

---

* The full list of official social media pages of the universities from this research can be seen at https://drive.google.com/file/d/1xBZGgQPuH2JN2_M7R-mPLMWC-ftQzPxg/view?usp=sharing

3 Results and discussions

3.1 Universities' SMM

The last group of moderating factors we examined is the universities' social media marketing policies. Therefore, we intended to check whether universities in the EaP countries integrate a proactive and analytics-based SMM as a part of their communication strategy in COVID-19 pandemic times.

H1: Universities fall behind in their SM activity: the dynamics of the firm-generated content (FGC) (number of posts) is lower than the dynamics of the customer engagement (ER).

To test this hypothesis for each university, we determine the average level of activity on both social networks (University posting rank). Accordingly, we also rank the activity of users (University ER rank). We show the results in Fig. 2. for each university and Fig. 3 for each EaP country.

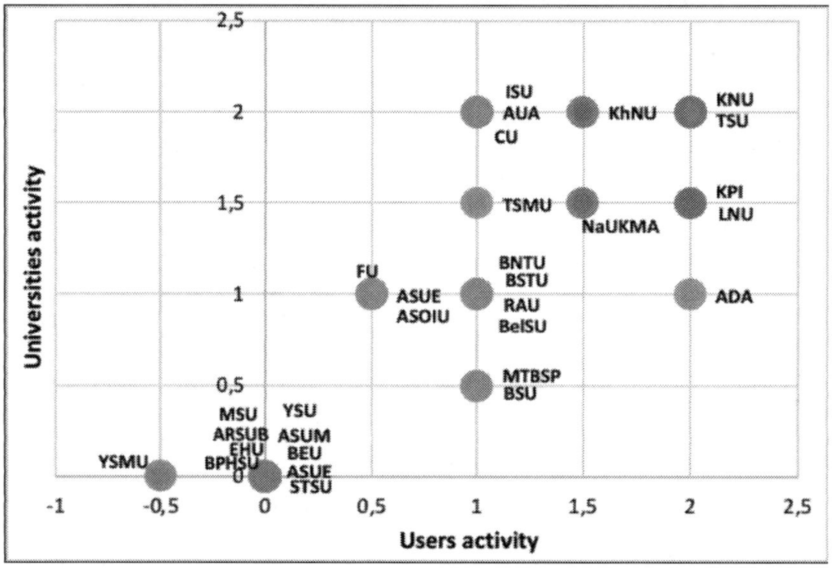

Fig. 2. The average level of activity for each university.

Thus, as can be seen from the matrix, only two universities have the best SM activity indicators (2; 2): TSU and KNU. They are constantly active, and their users are just as responsive. We call them the "VIP" segment.

The next group of universities belongs to the "Growing" segment. NaUKMA, LNU, KPI, KhNU - are educational institutions that demonstrate activity growth in at least one social network. Users, respectively, also responded adequately and increased interest in the proposed content.

The largest segment of "Newbies" includes 47% of the universities in question: RAU, ASUE, ISU, AUA, CU, TSMU, ADA, BSU, ASOIU, BelSU, MTBSP, BSTU, BNTU, FU. These educational institutions have not increased their activity in social networks, or vice versa - their users have not increased their interest in the proposed content.

 https://doi.org/10.1051/shsconf/202112608001

The worst situation is with the "Lost" segment - universities that do not have social network pages or closed their pages in the studied period. Unfortunately, 33% of the universities turned out to be in this segment.

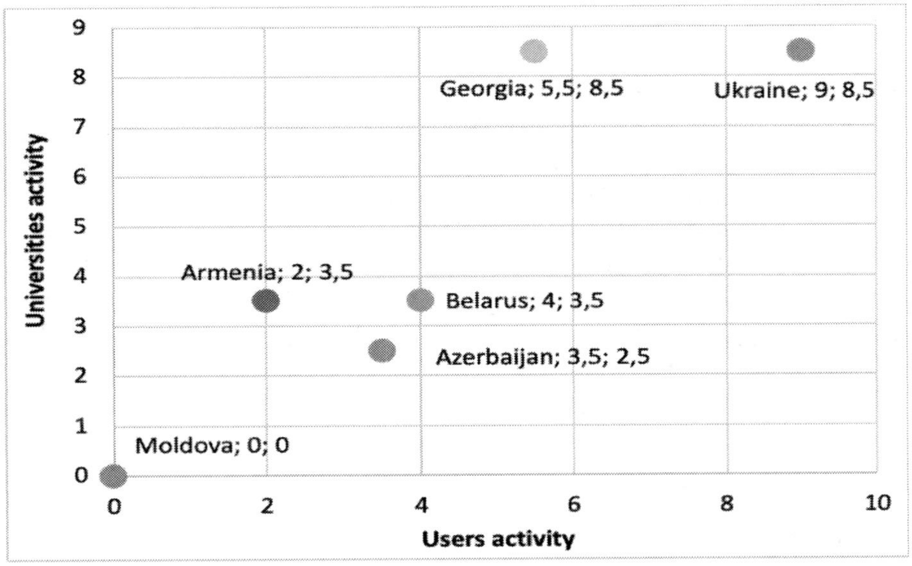

Fig. 3. The average level of activity for each EaP country.

As we can see, the universities of most countries (Fig. 3) did not set themselves the task of increasing the interaction with their stakeholders. However, in some educational institutions, the activity of the users themselves has increased. Thus, we can say that users want to receive news and other content from the university and are willing to actively respond by increasing the number of likes, comments, etc.

H2: Universities increase the content type according to the ER and stakeholders' preferences.

The next logical question is whether universities base their SM content decisions on users' reactions and engagement data. To test hypothesis H2, we selected seven universities that have increased activity in social networks from 2019 to 2020. These are TSU, KNU, LNU, NaUKMA, KPI, KhNU, ADA. First, we defined what type of content received the most significant response among users in 2019. We then checked whether educational institutions increased their activity relative using the most popular type of content in the COVID-19 period.

As can be seen, for Instagram, the results are pretty obvious (Fig. 4): Text and Photos have the highest ER. Moreover, in 2020, the universities we considered increased the production of this type of content. The Ukrainian institution KPI demonstrated extraordinary growth. It increased the usage of these content types 66 and 31 times, respectively. As can be seen, the leader in content growth in 2020 is video.

Nevertheless, not all institutions get an enthusiastic reaction to it from stakeholders. The results of the study for Facebook (Fig. 5) are not so noticeable. Each educational institution had a different dominating type of content. However, the most significant increase is seen in video posts. Moreover, although users have actively responded to all content types, some educational institutions reduced the number of posts of a specific type.

https://doi.org/10.1051/shsconf/202112608001

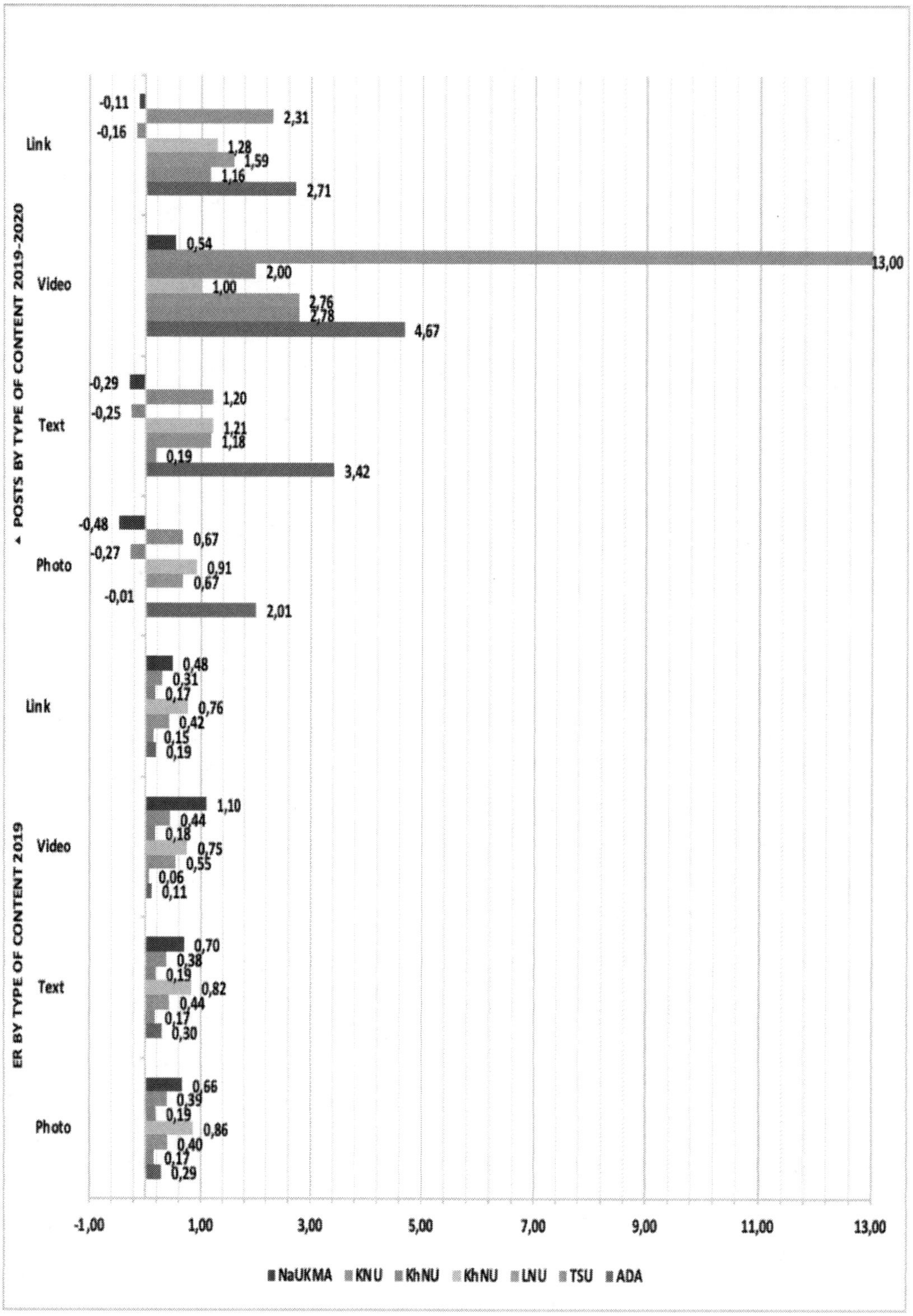

Fig. 4. Using analytics to manage content on Instagram.

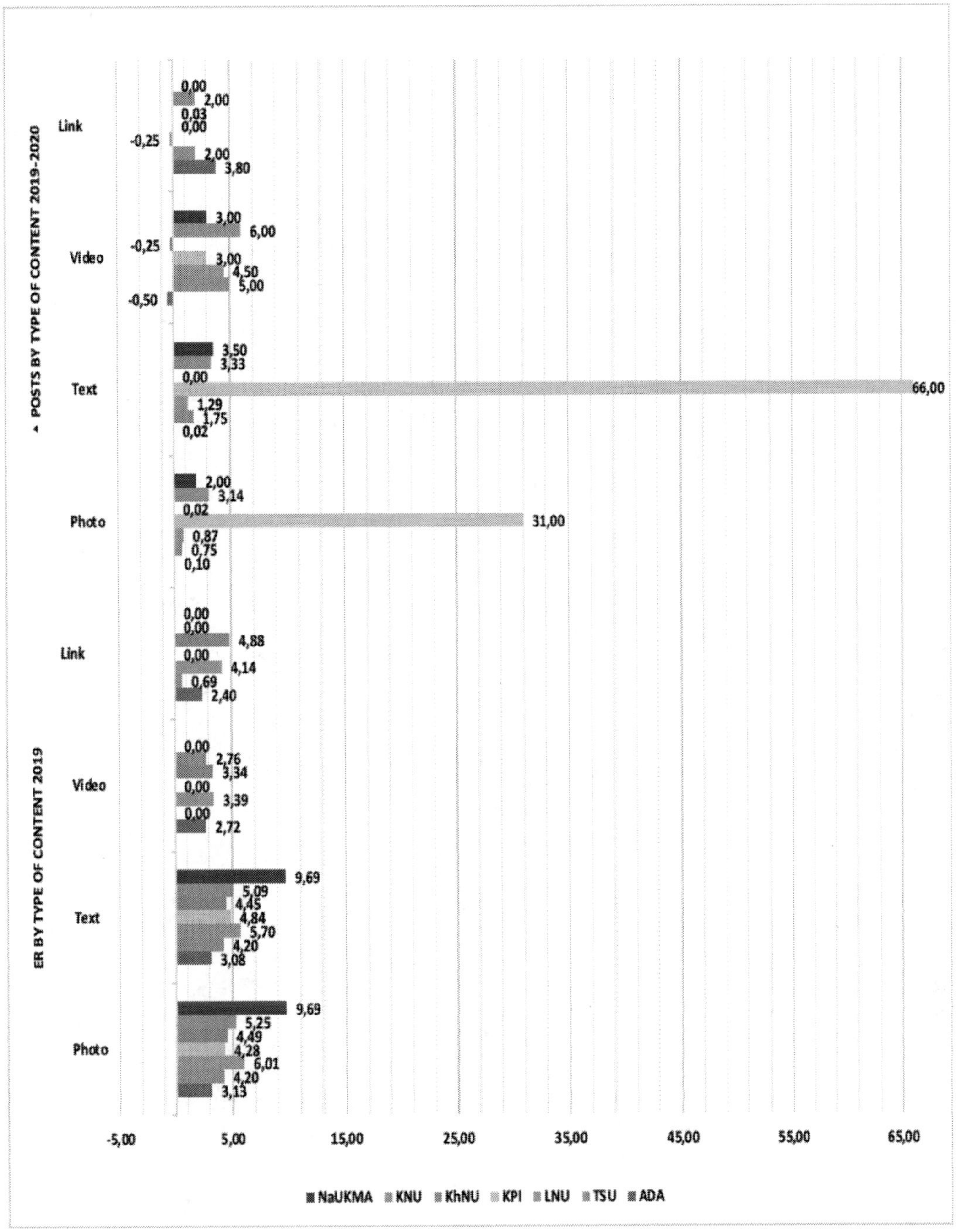

Fig. 5. Using analytics to manage content on Facebook.

3.2 Research implications

In this section, we outline opportunities for further research. A similar study needs to be done for an extended time frame, including 2021 and 2022 (when we talk about life after the COVID-19 pandemic and the new normal). The research can be based on the framework we presented. However, taking into account the results, researchers may eliminate some of the hypotheses. Below is the list of possible remaining assumptions for post-COVID-19 research:

- H1: Universities fall behind in their SM activity: the dynamics of the FGC (number of posts) is lower than the dynamics of the customer engagement (ER).
- H2: Universities increase the content type according to the ER and stakeholders' preferences.

Despite the pandemic slowing down, countries continue using restrictions that affect higher education. Therefore, it is necessary to follow which of the changes in the CM will remain and which will not have inertial development. Will the social activity of universities intensify in the post-COVID-19 period as a response to users' engagement? Will users demonstrate the same level of involvement? Or will both groups show high levels of "digital fatigue"?

Furthermore, a conceptual framework is needed to define certain places of different types of SM in the University managerial processes. Based on the best global practices, such framework will show educational institutions from EaP countries how to benefit from SM not only for communication purposes but also for improving value proposition (e.g., distance student mobility, professional mentoring and expertise), employment (e.g., LinkedIn or other professional networks), enrollee attraction (e.g., TikTok or new emerging SM for youth). This approach should concentrate not on the specific SM (as their popularity and penetration are different in different regions) but their function and target audiences.

Consequently, such a framework will imply an analytics-based approach to content creation. Future research papers must analyze the content and topics that elicit user responses. How crucial is the role of UGC (user-generated content) and FGC (firm-generated content) in the market of educational services? How to determine the optimal structure of UG and FG content. At what stage of the student's life cycle does the active participation with universities' SM terminate? Moreover, is it possible to extend the student's lifetime within the university scope using SM tools?

3.3 Managerial implications for universities

The process of managing profiles on social networks remains challenging to understand for university management. Our research shows that almost no university builds its communication on social networks based on analytical data. Analyzing the information offered by universities, we found that different types of content may predominate for various social networks, and, accordingly, users may prefer other activities from educational institutions. The choice of the content type is random rather than reasonable.

Currently, universities operate haphazardly without analyzing statistics on the preferences of their stakeholders. Such actions can negatively affect user interest and reduce the number of followers. By collecting and analyzing statistics on information preferences and responding accordingly, universities will increase the number of interested users and actively attract new ones by disseminating information about themselves in the world.

The same random approach is used when choosing the social networks themselves. Users have become more active across various social networks. Still, the universities of the Eastern Partnership countries are in the paradigm of "sales" thinking, when the main university product is the curriculum, and the primary target audience is the enrollee. The universities of the Eastern Partnership countries do not pay enough attention to the development of interaction with stakeholders in the digital space and social networks, as shown by the presented analysis. The scarce activity of universities in professional networks (Linkedin and others) indicates traditional employment assistance methods. The pull of the most used SM suggests a lack of understanding and segmentation of the target audience. Universities prefer Facebook, while recent statistics reveal, for instance, that Ukrainian students prefer Telegram and Instagram [17].

It is also worth noting that informal profiles of university communities, pages of individual departments, and divisions were left out of our attention. Nevertheless, such profiles exist. Moreover, they are a part of the process of building effective communication on social networks. Therefore, universities must decide on the level of decentralization of this process. On the one hand, decentralization will make content relevant to different target groups and provide higher ER.

However, as they increase their activity, universities must be prepared for positive and negative feedback, affecting the HEI's reputation [5, 18, 19]. Though some authors outline the possible negative market value effect from customer engagement initiatives for public companies, they clarify that less advertised companies (e.i. Universities) can benefit from it [20]. Accordingly, this confirms that universities should be active in social networks and collect and analyze statistics of their activities. Examine the comments, both positive and negative, because their consideration can increase the competitiveness of educational institutions not only in the local but also in the global market of educational services.

4 Conclusions

This paper focuses on how the online engagement of university stakeholders changed during the COVID-19 pandemic to meet their communication needs in the new reality. Our framework focuses on three factors modifying the COVID-19 influence on ER in six countries of the Eastern Partnership. The main factor that may have influenced such differences is the SMM policies of the Universities. We propose several further research directions, including the post-COVID-19 monitoring analyses and deeper content examination. As the research results reveal a favorable external environment for further customer engagement but a highly underestimated SM role inside the University management teams, we outline the number of managerial implications for universities.

References

1. K. Ihejirika, A. Goulding, Ph. Calvert, J. Libr. Adm., **61 (1),** 58-85 (2021).
2. R. Roopchund, V. Ramesh, V. C. Jaunky, *Use of Social Media for Improving Student Engagement at Université Des Mascareignes (UDM),* in Information Systems Design and Intelligent Applications. Proceedings of Fifth International Conference INDIA, Vol. 2 / [ed] S.C. Satapathy, V. Bhateja, R. Somanah, X.Sh. Yang, R. Senkerik (2018).
3. N. H. Al-Kumain, A. Alhazmi, T. Rmayah, M. S. Shabbir, N. Gazem (2021), *Sustaining Continuous Engagement in Value Co-creation Among Individuals in Universities Using Online Platforms: Role of Knowledge Self-efficacy, Commitment and Perceived Benefits*, Front. Psychol. (published electronically February 12), (2021).
4. O. Voropai, K. Pichyk, N. Chala, Economics and Sociology, **12 (4)**, 214-226 (2019).
5. V. Horodovenko, O. Bilovodska, V. Vatras, I. Kanzafarova, R. Melnyk, Estudios de Economia Aplicada, **9 (7),** Sp. Is.: Impact of Current Trends in Social Commerce, Economics, and Business Analytics (2021).
6. Office for National Statistics, Coronavirus and the Impact on Students in Higher Education in England: September to December 2020, London: ONS (2020).
7. V. Govindarajan, A. Srivastava, Harvard Bus. Rev. (March), (2020).
8. S. N. Mohd Anis, M. A. Ibrahim, L. Mohd Tahir, B. Abu, A. Khan, and R. A. Azizi, J. Adv. Res. Dyn. Contr. Syst., **12 (7)** (Special issue), 2929-2933 (2020).

9. M. Fenwick, J. A. McCahery, E. P.M. Vermeulen, EBOR, **22 (1),** 125-145 (2021).
10. B. Gesmen, I. Eken, *Students' Satisfaction with Applications Implemented in Distance Education Process During the Pandemic Period*, in *Management Strategies to Survive in a Competitive Environment.* Contributions to Management Science, H. Dincer, S. Yüksel, eds., Cham: Springer International Publishing, 75-89 (2021).
11. P. Escamilla-Fajardo, M. Alguacil, S. López-Carril, J. Hosp. Leis. Sport TO, **28** (2021)
12. Y. Zulnaidi, N. H. Md Saad, TEM Journal, **9 (4),** 1732-1739 (2020).
13. D. Tolkach, S. Pratt, J. Hosp. Leis. Sport TO, **28** (published electronically March 14), (2021).
14. The InfluenceRanking™ Engine: Our Ranking Technology Explained, *Academic Influence*, (accessed April 2021), (2021).
15. Complete Report Library, Datareportal, accessed June 14, 2021, https://datareportal.com/library (2021).
16. Engagement Rate: How to Calculate ER for Social Media Properly, Popsters, https://popsters.com/blog/post/er-how-to-calculate-engagement-rate (2017).
17. Attitude of the Population to the Media and Consumption of Different Types of Media in 2020, Research Report InMind August (2020).
18. S. Kvitka, G. Starushenko, V. Koval, H. Deforzh, O. Prokopenko, *Marketing of Ukrainian higher educational institutions representation based on modeling of webometrics ranking*. Marketing and Management of Innovations, **3**, 60-72 (2019).
19. O.V. Prokopenko, A.M. Karminsky, O.V. Klymenko, *The role of ratings in the educational process of higher education institutions*. Marketing and Management of Innovations, **4(2),** 141–146 (2011).
20. F.M. Beckers Sander, J. van Doorn, P.C. Verhoef, *Good, Better, Engaged? The Effect of Company-Initiated Customer Engagement Behavior on Shareholder Value*, J. Acad. Market. Sci., **46,** 366-383 (2018).

Impact of the Covid-19 Pandemic on Entrepreneurial Capacity and Financial Security in Ukraine

Yana Derbenova[1]*, Ganna Kashyna[2], *Serhii* Leontovych[3], *Olena* Ablova[4], *Oksana* Dmitrieva[5], *and Lyudmyla* Yaremenko[6]

[1]Taras Taras Shevchenko National University of Kyiv, 01033, str Volodymyrska 64/13, Kyiv, Ukraine
[2]Institute for Military and Strategic Studies of the National Defense University of Ukraine named after Ivan Cherniakhovskyi, 03049, Povitroflotsky Avenue, 28, Kyiv, Ukraine
[3]Center for Military and Strategic Studies of the National Defense University of Ukraine named after Ivan Cherniakhovskyi, Kyiv, Ukraine
[4]WC2R, Great Britain, London
[5]Kyiv National Economic University named after Vadim Hetman, Victory Avenue, 54/1, Kyiv, 03057 Kyiv, Ukraine
[6]Central Ukrainian State Pedagogical University named after Volodymyr Vynnychenko, Departments of Applied Mathematics, Statistics and Economics, 25000, str. Shevshenko 1, Kropyvnytsky, Ukraine

Abstract. The article analyzes the impact of the COVID-19 pandemic on business activities in Ukraine and the world and its financial security. The analysis showed that the economy of Ukraine and most countries of the world suffered significant losses from the introduction of anti-epidemic restrictions on business activity, which turned out to be in the fall of GDP, index of industrial production investment activity. Because of temporary or complete shutdown of business the unemployment rate increased significantly. The pandemic crisis also negatively impacted the financial results of enterprises, which decreased significantly in all kinds of economic activities, especially in industry, transport, temporary accommodation and catering, where the share of unprofitable enterprises reaches 73%. Thus, the problem of ensuring the financial security of business becomes relevant. The aim of the work is to develop methodological approaches to assessing the impact of the pandemic crisis on the financial security of business and to identify effective ways of ensuring such security. In the process of research methods of scientific abstraction, comparative and system analysis and synthesis, systematization and logical generalization were used. It was determined that the pandemic affected small and medium businesses most of all, where a strong negative impact was experienced by more than 60% of entrepreneurs. Large businesses were less affected, but they also suffered significant losses due to the decline in business activity and demand in the domestic and foreign markets. Agriculture was least affected by the pandemic, which suffered the least losses, primarily due to the insignificant impact of restrictions affecting the activities of this business. The paper systematized the risks arising from the factors of the pandemic crisis,

* Corresponding author: yanaderbenova@ukr.net

identifying their manifestations and financial implications for business. As shown by the research, such consequences are manifested in the form of full or partial loss of profits, problems with debt repayment, reduction of financial stability, liquidity, increase in accounts receivable, increased costs for anti-epidemic measures, etc. Thus, the procedure for assessing the business's financial security is to take into account all factors that shape the financial situation of business entities during a pandemic crisis in a comprehensive manner. Keywords: financial security, pandemic crisis, risks, financial stability, security indicators, business financial performance.

Introduction

An unexpected problem faced by humanity in early 2020 was the COVID-19 pandemic. The pandemic had a significant impact on business activities worldwide and in Ukraine in particular. Especially negative were the consequences of the so-called "lockdown" Lockdown imposed in the spring by most European countries, and Ukraine as well. The complete suspension of work of the majority of trading, service enterprises and transport companies entailed a wave of bankruptcies. Also the unemployment increased and the level of unprofitability of business increased. All of these consequences led to a drop in GDP and decreased revenues to budgets of all levels. The investment processes were curtailed.

The pandemic crisis has become a challenge for most businesses. Therefore, issues of ensuring their financial security have come to the forefront. Financial security is the ability to maintain a positive financial result of the enterprise, the preservation of the value of the enterprise.

Literature review

In 2020, the COVID-19 pandemic sparked the emergence of scientific research. The studies are related to the impact of this factor on the financial security of the business. In this respect, it is necessary to note the work of N. Fedulova and M. Dzhulai. [1]. The authors use a detailed analysis of the risks faced by businesses in recent years in Ukraine. They offer ways to improve the safety of business from the position of preservation of financial stability.

Also noteworthy is the work of Ivanyuk [2], where he conducts a systematization of factors for the formation of a favourable business environment and business security. The study identifies the main reference points of state support for entrepreneurial activity in the increasing influence of global pandemic challenges.

Strategic priorities of formation of economic security in pandemic conditions at different levels of economic system are considered in their work V. Gorbulin and Y. Danik [3]. This list is far from being complete, with the development of the pandemic there are constantly new publications and research results of the search for effective directions of business security in these conditions.

The purpose of this study is to develop methodological approaches to assessing the impact of the pandemic crisis on the financial security of business and determine effective ways to ensure such security. Due to the crisis situation caused by pandemic COVID - 19, Ukrainian business in 2020 found itself on the verge of survival. Economic recession, reduction in purchasing power and changes in the behavior of citizens, the lack of support (or its ineffectiveness) from the state led to a reduction in production, or even the shutdown of entire sectors of the economy.

Results

During the crisis, the government introduced a strict regulatory policy, which created barriers to work, reducing the profitability of business. Small and medium-sized businesses were the first to be threatened. Large corporations were also forced to downsize, sending employees on free vacations. Overcoming the consequences of the coronavirus pandemic and quarantine restrictions on the Ukrainian economy is impossible without taking into account the vision and needs of the beneficiaries themselves - business representatives.

The State Statistics Service of Ukraine in 2020 reported that the fall in GDP (percent to the corresponding quarter last year) represented in the first quarter - 1.3%, in the second - 11.4%, in the third - 3.5%. At the same time, the largest fall was experienced by such types of economic activity as: temporary accommodation and catering, transport, professional, scientific and technical activities, agriculture, provision of other types of services. The index of industrial production experienced a significant decrease, the dynamics of changes in which is shown in Figure 1.

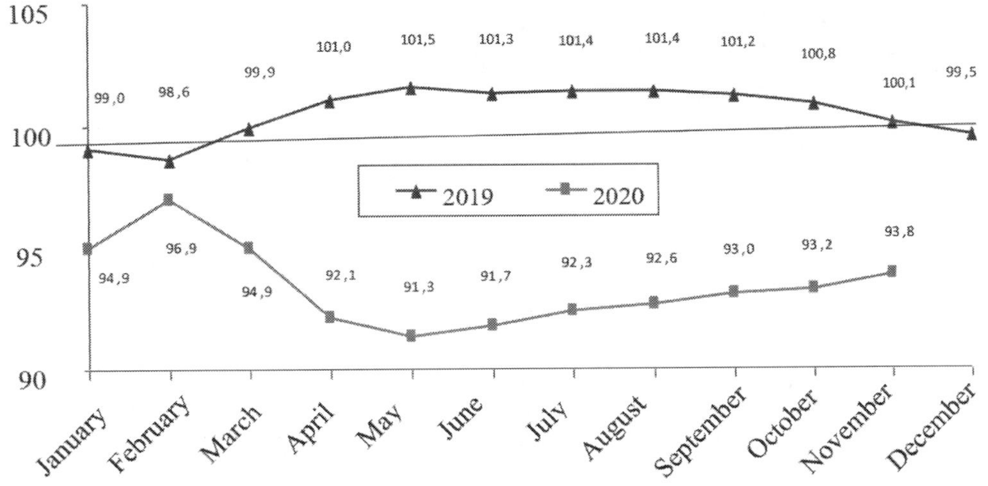

Fig. 1. Indices of industrial production (in % to the corresponding period of the previous year, cumulative total) [5].

From January to September 2020 the volume of capital investments decreased significantly. As compared with the same period of the previous year they represented 64.6%. The key source of investment remains own funds of enterprises and organizations - 69.2%, which have experienced a significant decrease as a result of the factors of the pandemic crisis.

The introduction of quarantine measures and a decrease in the activity of many entrepreneurial strictures, especially in small business, led to a sharp increase in the level of unemployment in Ukraine. The number of unemployed rose in 2020 from an official 349,000 persons in March to 457,000 in April, 511,000 in May, and 517,000 in June. And only after some mitigation of quarantine restrictions, the number of unemployed decreased to 506 thousand in July and to 474 thousand in August. The slight improvement is partly due to renewed activity of labor migrants.

With the beginning of the pandemic, the Ukrainian financial market was also negatively affected. Figure 2 shows the dynamics of the Ukrainian Equity Index (UX) according to the Ukrainian Exchange [6]. As can be seen from the graph, the UX index experienced a significant drop precisely with the onset of the pandemic crisis in March 2020 and its gradual resumption began only at the end of October.

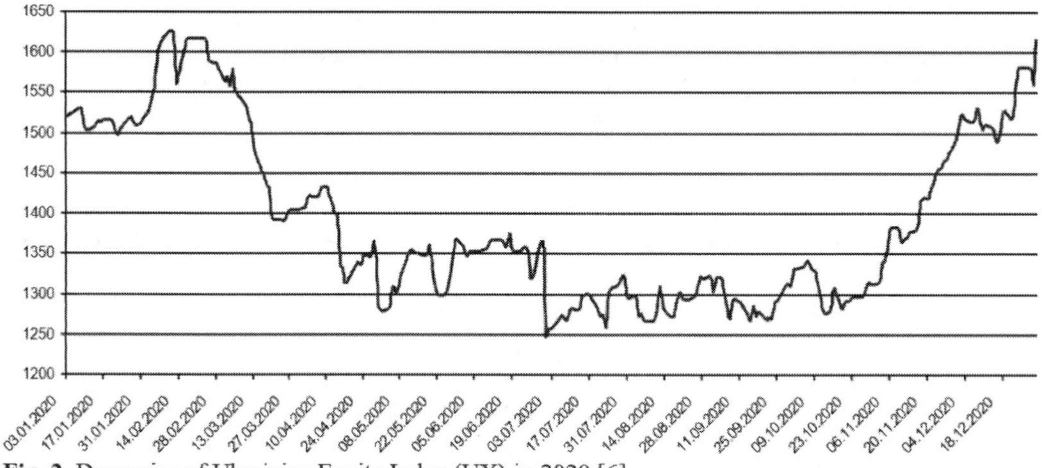

Fig. 2. Dynamics of Ukrainian Equity Index (UX) in 2020 [6]

The negative impact of the pandemic crisis led to a deterioration in the financial results of enterprises in almost all spheres of economic activity (Table 1). This was especially evident in industry, transport, professional scientific and technical activities, art and sports.

Table 1. Financial results of enterprises [5]

Types of economic activity	Financial result to taxation		The number of unprofitable enterprises in % of the total number of enterprises	
	January-September 2019, mln hryvnas	January-September 2020, mln hryvnas	January-September 2019	January-September 2020
1	2	3	4	5
Total	342809,5	93318,0	22,7	34,8
Agriculture, Forestry and Fisheries	-43,3	-92,8	50,0	35,7
Industry	155395,9	-44856,9	26,1	36,9
Construction	3638,3	1628,9	21,1	27,0
Wholesale and retail trade; repair of motor vehicles and motorcycles	65600,0	12512,7	16,2	28,1
Transport, warehousing, postal and courier activities	8171,9	-14888,3	31,4	54,1
Temporary accommodation and catering	2603,8	-2889,1	18,2	73,2
Information and telecommunications	16940,9	10286,9	15,1	30,0

 https://doi.org/10.1051/shsconf/202112608002

Table 1. (continued)

1	2	3	4	5
Financial and insurance activities	46268,6	165140,7	14,3	13,6
Real estate operations	11020,6	-7895,8	29,5	57,9
Professional, scientific and technical activities	30005,6	-24251,5	29,5	34,0
Administrative and auxiliary service activities	2300,5	116,1	29,9	38,5
Education	-2,7	-67,9	33,3	33,3
Health care and social assistance	378,6	835,9	22,6	32,6
Arts, sports, entertainment and recreation	416,9	-2260,4	40,0	63,3
Provision of other types of services	113,9	-0,5	-	25,0

The pandemic crisis has affected almost all spheres of entrepreneurial activity, especially affecting small and medium businesses. Such sectors as transport, manufacturing, construction, wholesale and retail trade, accommodation and catering services, real estate operations, professional services and other services for the population (e.g. hairdressing services) suffered the most. This is confirmed by a report of the International Trade Center (ITC) entitled "COVID-19 - The big isolation and its impact on small business. The results of a survey conducted by ITC among different companies about the impact they faced through the crisis COVID - 19, is shown in Figure 3.

Similar surveys conducted by INFO SAPINS LLC (limited liability company) for the Representative Office of the Center for International Private Enterprise in Ukraine among Ukrainian small and medium businessmen [8] demonstrated that the main obstacle during the pandemic is the limitation of business activity which together with the reduction of demand at the domestic market and low purchasing power has the most negative effect on business activity. At the same time, economic entities of all sizes are equally affected by this factor: close to 70% of respondents assessed the impact as "very negative" or "negative". On the whole, quarantine measures, which consist in the introduction of restrictions in the work of businesses led to an overall negative impact on their situation (Fig. 4).

As it can be seen from the diagram, the second most important factor is the continuation of quarantine measures, which has a negative influence on, respectively 69% of micro-, 65% of small and 58% of medium enterprises respectively. Also about half of the respondents indicated difficulties with logistics as a negative factor. One in three small and medium-sized businesses experienced problems through disruption of supply chains and the need to maintain social distance in the communication process. Also influential for exporters was a decrease in demand at international markets. The problems faced by business in connection with the coronavirus pandemic manifested themselves in the emergence of new risks and challenges to which most business structures were not prepared. Especially vulnerable to them were business entities in the sphere of small and medium business, which, according to the results of surveys [8], incurred significant financial losses (Fig. 5).

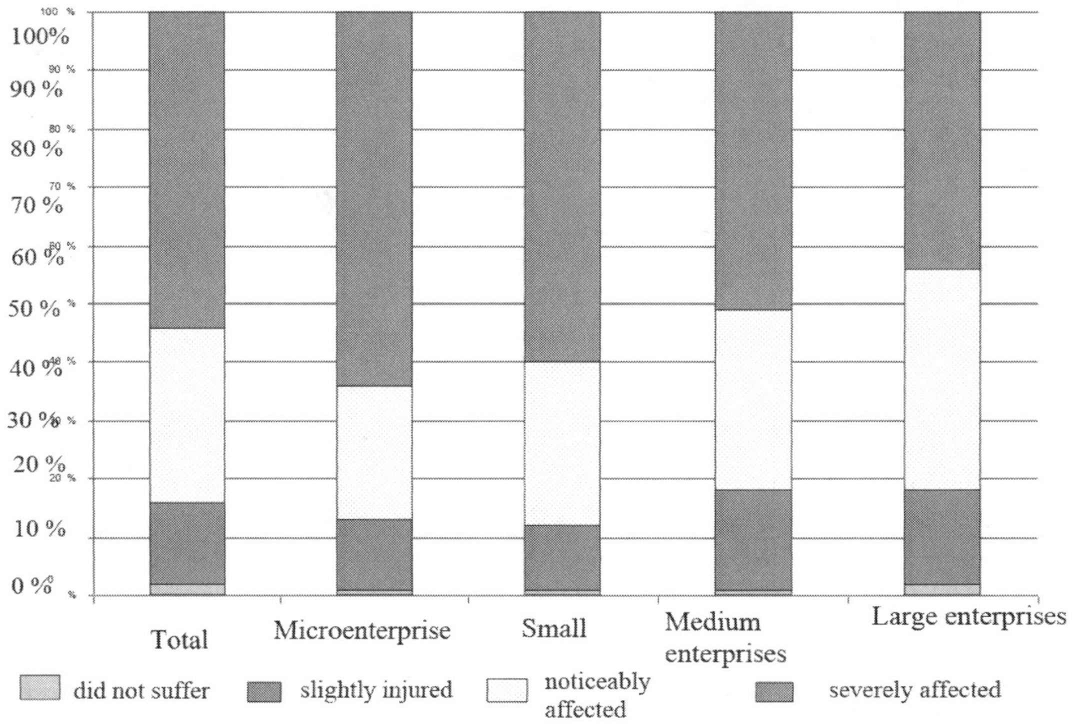

Fig. 3. Impact of the pandemic on the business sphere depending on the scale of the business [7]

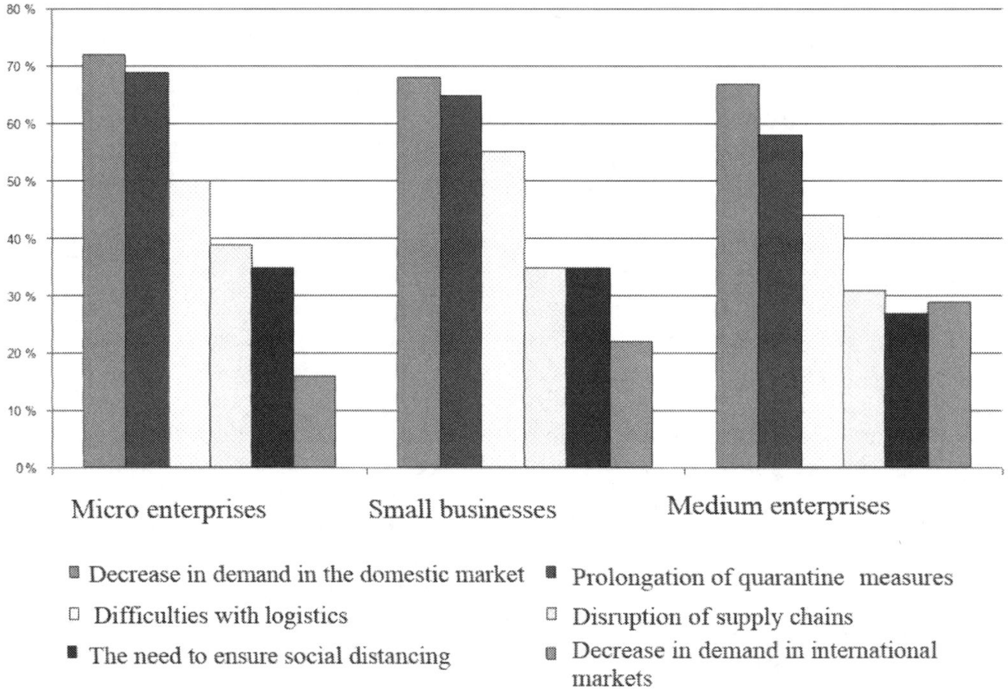

Fig.4. Impact of pandemic restrictions on business operations in Ukraine [8]

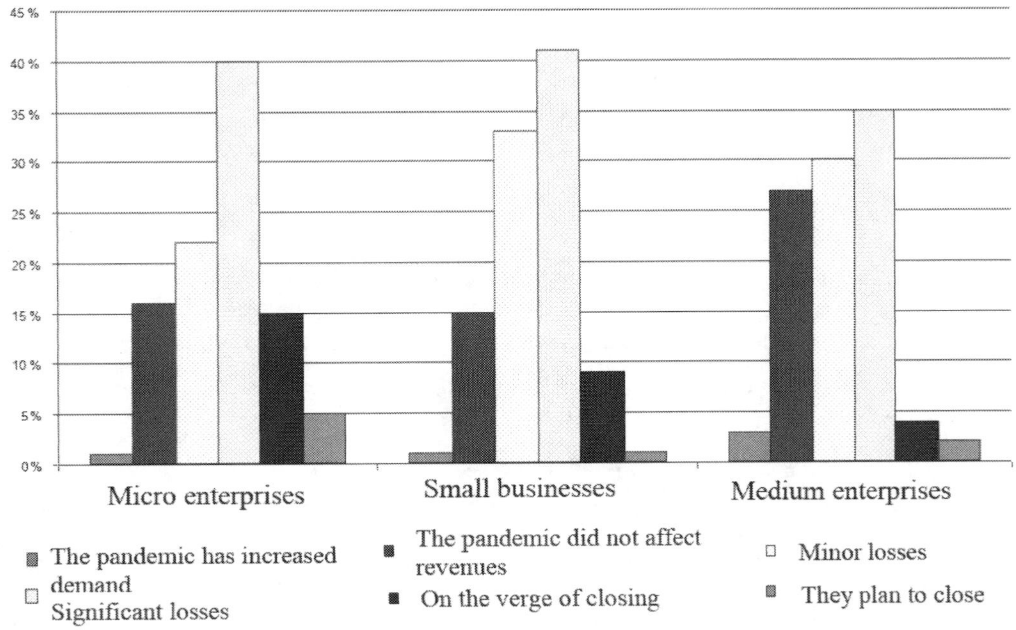

Fig. 5. Business losses from pandemic and quarantine measures [8]

As can be seen from the graph, medium-sized enterprises are the most successful in overcoming the difficulties of the pandemic situation, for 3% of which demand for products and services has even increased, and for 27% of whose revenues have not suffered losses from the pandemic.

The analysis (Fig. 6) showed that such sectors as trade, industry, leasing, logistics, transport suffered the most from the pandemic, suffering the greatest losses. Agriculture was least affected by the pandemic, which incurred the least losses, primarily through the insignificant impact of restrictions that affected the activities of this business.

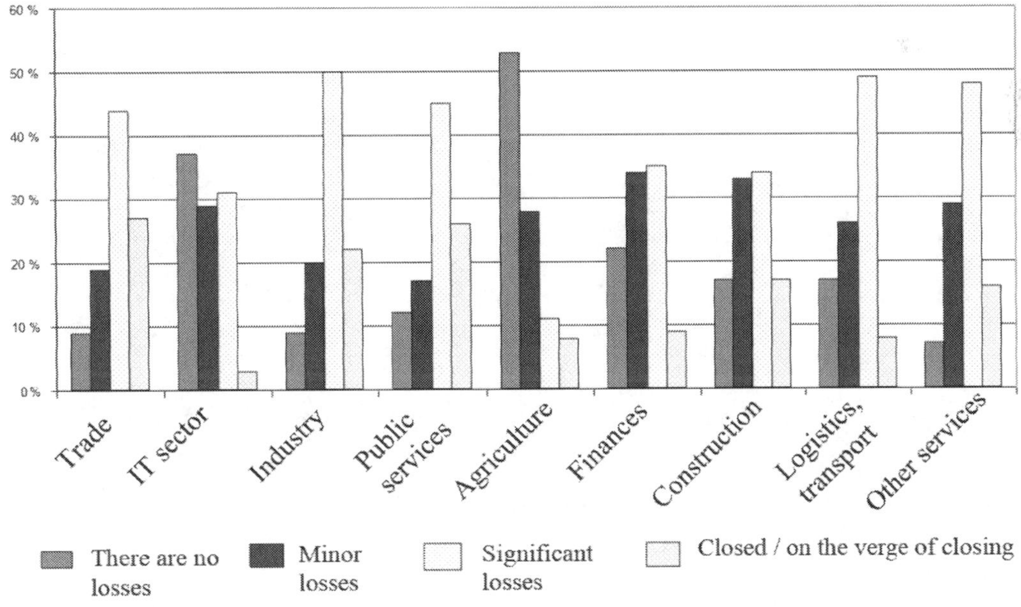

Fig. 6. Pandemic Business Losses by Activity [8]

Many enterprises have problems with liquidity and solvency, as for more than 2 months there were no sales. Accordingly, there are difficulties with repayment of accounts payable, taxes, salaries, rent, loans and so on. In addition, despite the removal of some restrictive procedures, sales are also weak. In this connection, the state pursues a policy of supporting business by providing, on the one hand, very attractive loans and, on the other hand, by stimulating solvent demand

The results of the research show that the financial security during business pandemic is affected by external (exogenous) and internal (endogenous) risks that arise during the activities of business entities. The action of the corresponding risks leads to a decrease in the financial results and indicators that characterize the financial condition of the enterprise and the level of its financial security. The systematization of such risks is shown in Table 2.

Table 2. Pandemic risks and their financial implications

Source of occurrence	Manifestation of risk	Financial consequences
Exogenous risks		
Restriction or complete prohibition of activity	Reduction or complete cessation of production of products or provision of services. Release of employees. Interruption of investment and innovation processes. Loss of sales markets, connections with suppliers and buyers. Inability to obtain credit	Partial or complete loss of profit, reduction of profitability, financial stability, liquidity. Problems with debt repayment, possibility of bankruptcy
Reduction of demand on domestic and foreign markets	Decrease in the volume of sales of products. Reduction of employees. Loss of markets, connections with suppliers and buyers. Formation of excessive stocks of finished goods.	Decrease in revenues and profits. Deterioration of profitability indicators. Increase in expenses for finished product storage and marketing activities
Restrictions in transport operations	Disruptions in the supply of raw materials and shipment of products. Complications of logistic processes. Difficulties with the implementation of foreign economic operations	Increase in accounts receivable, increase in expenses for logistics activities and delivery of employees. Decrease in profits. Decrease of currency receipts
Endogenous risks		
Illness of workers at COVID - 19	Loss of human resources potential. Decrease in labor productivity. The need to search for staff replacements.	Decrease in revenue and profits. Sick pay expenses. Additional expenses for employee recuperation
Necessity to comply with anti-epidemic requirements	Need for protective equipment, medical equipment (thermometers, disinfectors, etc.). Implementation of anti-epidemic measures	Additional expenses for anti-epidemic measures, which reduce the total financial result

Theory and practice appeals to a huge number of a huge number of indicators to assess financial security, which complicates to some extent the implementation of the analysis, and often gives different results in general. This leads to difficulties in comparing the results of different standards when forming conclusions about financial security. Research shows, that most of the existing methods of financial security analysis contain different, but at the same time somewhat similar methodological principles of such evaluation, the choice of which should take into account the specifics of entrepreneurial activity. Existing criteria of evaluating the level of financial security should be adjusted in order to determine their marginal values. At the same time, it should be borne in mind that quite often one of the peculiarities of entrepreneurial activity is a combination of interests of different categories of recipients of information, since the owner may act both as an investor and a manager.

For the business owner an important characteristic is the return on invested capital in the current period and in the long run. At the same time, the main goal for him is the growth of the value characteristics of the business, which corresponds to the concept of VBM. At the same time managers of the enterprise are mainly interested in factors that affect the performance of current activities, the use of human resources, material and other resources, the development of the business as a whole and the achievement of financial results.

Conclusions

Thus, it can be noted that in Ukraine as well as all over the world pandemic COVID19 significantly affected the activities of businesses of different forms and sizes. The most painful consequences it had on small and medium-sized businesses, most of which experienced certain restrictions and prohibitions, especially during the lockdown period. Financial results declined significantly in industry, transport, professional, scientific and technical activities, arts and sports. For these and other industries, the problem of ensuring the financial security of business came to the fore. The analysis showed that among the factors of pandemic impact on financial security of business, the most important are risks associated with restriction or complete ban on activities, reduction of demand in domestic and foreign markets, restriction in transport operation, illness of employees on COVID - 19.The consequences of the impact of such risks, there is a partial or complete loss of profits, reduced financial stability, an increase in accounts receivable, additional costs for anti-epidemic measures, which ultimately reduces the financial security of the business.

References

1. I. Fedulova, M. Julai, Bulletin of KNTEU, **4**, 74–91 (2020).
2. B. Ivanyuk, Financial space, **3 (39)**, 183-190 (2020)
3. V. Gorbulin, Y. Danyk, Bulletin of the National Academy of Sciences of Ukraine, **5**, 3-18 (2020)
4. *The Covid-19 pandemic plunges the world economy into a recession record. Le Monde*: Website. URL: https://www.lemonde.fr/economie/article/2020/09/01/la-pandemie-de-covid-19-plonge-l-economie-mondialedans-une-recession-record_6050618_3234.html
5. Official site of the State Statistics Service of Ukraine URL: http://www.ukrstat.gov.ua
6. Site of the Ukrainian Stock Exchange. URL: http://www.ux.ua/ua/
7. *International Trade Center. SME Competitiveness Outlook 2020: COVID-19: The Great Lockdown and its Impact on Small Business*. ITC, Geneva. URL:

https://www.intracen.org/uploadedFiles/intracenorg/Content/Publications/ITCSMECO2020.pdf

8. *Report on the study of micro, small and medium enterprises* URL: https://sapiens.com.ua/publications/socpol-research/135/Info%20Sapiens_CIPE_report.pdf

Financial Flexibility of Universities in Adaptation to Covid-19 Challenges

Liudmyla Yurchyshena[1*], *and Olexandra* Laktionova[1]

[1]Vasyl' Stus Donetsk National University, Vinnytsia, Ukraine

Abstract. During the adaptation of universities to Covid-19, the problem of their financial flexibility has become crucial. The ability to react quickly during a pandemic crisis with changes of training format and educational activities, the flexibility to respond to external challenges under the scarce resources, maintenance of a competitive position on the market of educational services are the symptoms and implications of financial flexibility. The purpose of the article is to determine the signs of the university's financial flexibility, which could be the basis for designing and practical application of methodological tools for evaluating the financial flexibility of universities. The concept of financial flexibility of universities is proposed as the ability of university management to generate income, restrain and optimally allocate costs, generate cash flows with constrained budget funding during the crisis, create and implement business models, implement unique valuable educational proposals. Manifestations of financial flexibility concerning state and private financing have been defined. The signs of financial flexibility, such as sensitivity, efficiency, manoeuvrability, alternativeness, economy, have been outlined. Methodical tools and indicators for the evaluation of financial flexibility have been proposed. The degree of the impact of main factors on the financial flexibility of universities is determined. Keywords: financial flexibility, Universities, sensitivity, efficiency, manoeuvrability, alternativeness, economy.

Introduction

The Covid-19 pandemic has become a challenge for higher education institutions not only in terms of rapid response to changes in the format of education but also a flexible response to external challenges, given the limited opportunities for quality educational activities, the ability to create and implement value proposals, work online, service campuses. The financial consequences became noticeable not immediately but gradually, along with the loss of universities' additional income, reduction of budget allocations, reduction of the solvency of students and entrants, reduction of the number of budget places. The consequences of the pandemic for free economic zones have become a stress test for flexible response to the situation, including financial. For the first time, free economic zones have been left alone with financial problems related to revenue generation and cost containment and optimisation,

* Corresponding author: author@email.org

changes in the university funding model, and the ability to be financially flexible in adapting to the new operation model.

Exploring the concept of «financial flexibility», Laktionova defines it as an appropriate response of the company to the manifestation of financial constraints, which has a dynamic nature, associated with the availability of capital from external sources, an adaptation of financial potential of economic entities to organisational change, management tools, financial relations [1].

According to Ang and Smedema [2] and Gamba and Triantis [3], financial flexibility is manifested in the ability of the institution to manage revenues during their reductions or fluctuations, get a positive effect from shocks, in particular, reduce external financing costs, invest, accelerate capital turnover. Ang and Smedema note the regularity that firms that can self-finance during the recession in conditions of limited access to external capital in the future will receive priorities for profitable investment [2].

Yaw M. Mensah, Robert Werner [4] determine that the efficiency of higher education institutions slows down in the absence of financial flexibility, and the degree of financial flexibility is assessed as the ratio of net assets to total assets. Researchers argue that the greater the financial constraints, the more efficiently the costs are distributed.

Scholars of China, Md Rashidul Islam, Man Wang, Leo Vashkor Dewri [5] synthesising approaches to the concept of "financial flexibility" identify it as an essential element in deciding on the capital structure, recognise the effect of financial constraints, the relationship of financial flexibility and performance in times of crisis, the relationship of financial flexibility, corporate governance and corporate social responsibility and in general, arguing for the positive effect of financial flexibility in a crisis.

The value of financial flexibility in the Covid-19 period is studied by Swiss scientists Rüdiger Fahlenbrach, Kevin Rageth, René M Stulz, arguing that firms with higher financial flexibility are better able to cover cash flow deficits due to the crisis, differentiate revenues, bear lower financial losses. As a result, financial flexibility forms certain competitive advantages of the company and their ability to be more stress-resistant to shock phenomena [6].

Thus, the financial flexibility of universities is determined by the ability of university management to generate income, restrain and optimally allocate costs, generate financial flows with limited budget funding during the crisis, create and implement business models, implement unique value proposals as a result of adapting universities to new realities.

Currently, the problem of financial flexibility of universities in terms of adaptation to the challenges of Covid-19 has been studied in terms of the main trends due to these processes, but further research requires methodological tools for assessing financial flexibility and their practical application on a sample of universities.

The purpose of the article is to determine the signs of financial flexibility of the university, development, and practical application of methodological tools to assess the financial flexibility of universities.

Results

The expansion of the financial autonomy of higher education institutions poses certain challenges for university management during the period of adaptation to Covid-19. The activity of universities is realised at the expense of budgetary and private financing, resulting in a generation of financial flows.

Financial flexibility in terms of adaptation to Covid-19 has different manifestations (Fig. 1), on the one hand, universities receive limited budget funding, which is a stable source of income. However, in most cases tied to the contingent and results of free economic activities, manifestations of financial Flexibility is restraint and optimal cost allocation, as the impact of universities on the growth of budget funding is limited. On the other hand, the growth of private funding is an indicator of the competitiveness of free economic resources, which can generate its income through the formation of value proposals, image programs, interaction with the business environment, rapid response and the university's ability to adapt to crisis conditions.

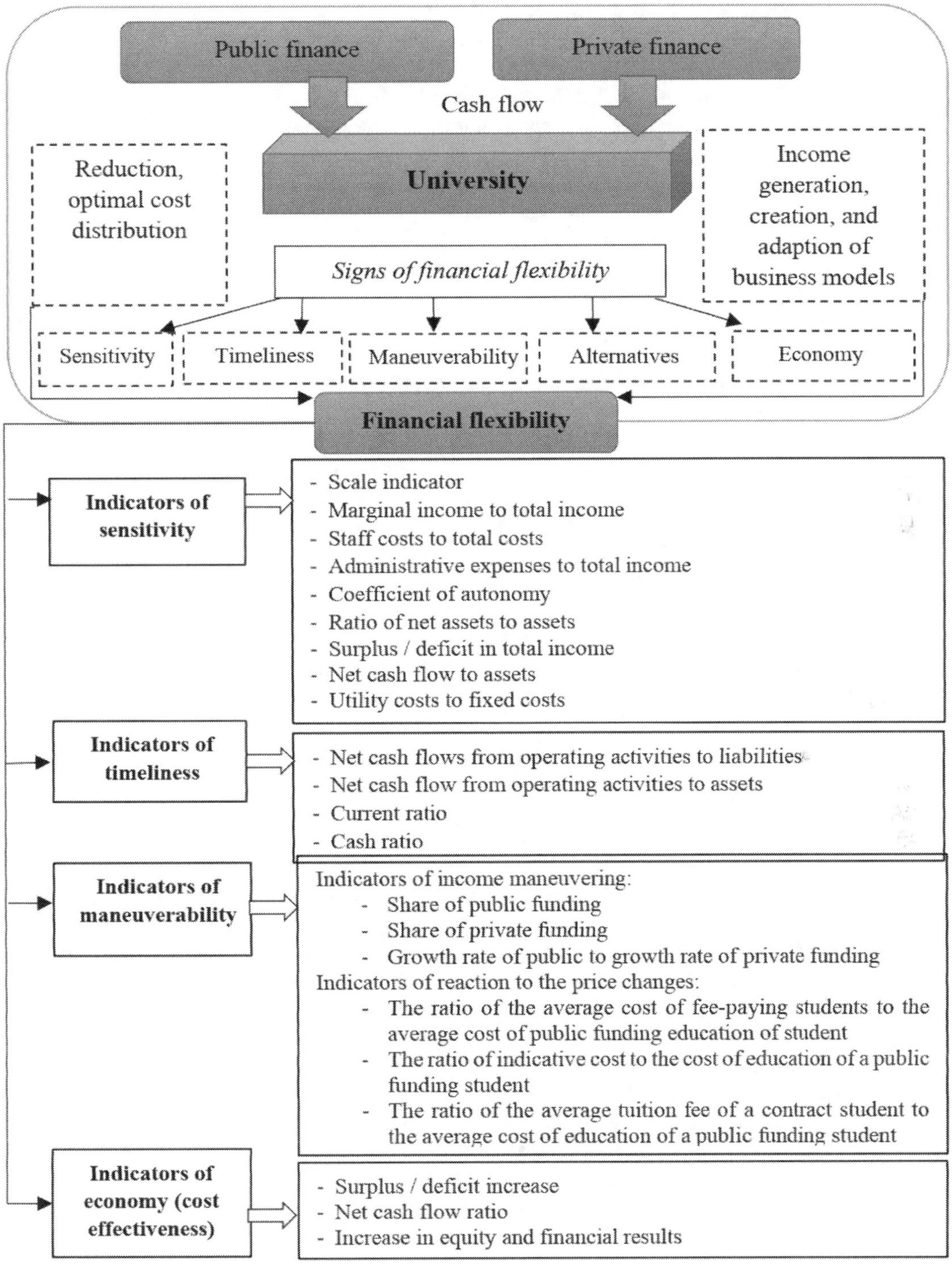

Fig. 1. Tools for evaluation of financial flexibility of university in terms of Covid-19

From the above, we can distinguish the following signs of financial flexibility:

- sensitivity shows how universities open to a crisis influences, have reserves in financial and performance conditions and so could be flexible in the period of adaptation to the negative outcomes of Covid-19, such as changes in the contingent, reduction of special fund revenues, availability of accumulated reserves, change in cost structure, redistribution, readiness to increase certain cost items;
- efficiency shows how quickly universities respond to change and are able to produce activities in the conditions of Covid-19;
- manoeuvrability characterises the availability of free economic resources for responding to changes, quickly adaptation to new funding conditions, changing the structure of funding sources by increasing private funding, implementation new progressive value proposals;
- alternatives are manifested in the ability of university management to make alternative financial decisions, to achieve specific strategic and tactical goals through various combinations of decisions and the production of alternative activities;
- efficiency is a resultant sign of financial flexibility, which shows the cost of the university's response, which is manifested in financial results, the ability to meet financial obligations on time, the presence of positive net cash flow, the ability to accumulate funds in a crisis.

For evaluation of financial flexibility, the system of indicators could be divided into categories:

1. Sensitivity indicators characterise universities' degree of sensitivity to changes and consequences of the impact of Covid-19. This group includes the following indicators:

- the contingent of students characterises the scale indicator, and the degree of sensitivity is determined by the size of the contingent, its structure (the predominance of public funding students make it more resistant to change) and dynamics. The positive trend in the indicator reduces the level of sensitivity to the crisis caused by Covid-19;
- marginal income to total income is defined as the ratio of the difference between income, labor costs, deductions for social activities, and depreciation to the university's total income. The higher this indicator, the greater the degree of sensitivity of the university to changes in income;
- staff costs to total costs, the more significant the share, the greater the degree of sensitivity and the lower the level of flexibility, and vice versa. Despite the fact that such costs are variable or dependent on the contingent of students, reduction of such costs in the educational sphere is challenging, since teaching and research personnel constitute the capital and the main value of universities;
- administrative expenses to total income – the smaller this one, the more resilient universities are to change. High administrative costs lead to an increase in the sensitivity of the institution to fluctuations in income;
- coefficient of autonomy is calculated as the ratio of equity to assets. The higher the value of the indicator, the less sensitive a university to crisis phenomena;
- the ratio of net assets to assets is calculated as the ratio of assets fewer liabilities, provisions and deferred income to total assets of a university. The decrease in this indicator in the dynamics increases the level of sensitivity of universities to change, affects the ability to meet financial obligations to counterparties;
- surplus/deficit increase – a decrease in the dynamics or negative value is an indicator of sensitivity and is negatively reflected in the financial results of the university;

- net cash flow to assets shows the ability of universities to form a financial cushion, the greater the value of the indicator, the more flexible the institution is, while the positive growth of the indicator in the dynamics and the value is greater than zero;
- the share of utility costs to fixed costs - the greater the share of utility costs, the greater the potential for universities savings in the period of adaptation to Covid-19, subject to the transition to online learning. All other things being equal, this indicator reduces the level of flexibility of the institution in the period of declining revenues and optimising the cost structure.

2. Indicators of timeliness show the ability of universities to finance current needs, to fulfil current financial obligations to counterparties in the face of declining revenues. Such indicators should include:

- net cash flow from operating activities to liabilities shows the level of ability of universities to meet current and long-term obligations; a positive value is one of the indicators of financial flexibility of the institution;
- net cash flow from operating activities to assets – the recommended value is greater than zero;
- the current ratio is calculated as the ratio of financial assets (net of long-term receivables and long-term financial investments) to current liabilities;
- cash ratio is calculated as the ratio of cash and cash equivalents of managers of budget funds and state trust funds, budget funds and other clients to current liabilities.

3. Indicators of maneuverability characterise the ability of universities to adapt to change, and include:

3.1. Indicators of income manoeuvring:

- share of public funding - the excess of the share of state funding over the average for universities in the industry is a sign of manoeuvrability, because public funding is a stable source of income to maintain the current financial stability of the university;
- share of private funding - the growth of such financing is an indicator of the competitiveness of universities on the one hand; on the other, there is a risk of dependence of private financing on Covid-19. Thus, most institutions in the period of adaptation to the pandemic reduce training costs, giving priority to funding other costs, the same trends apply to the private sector;
- growth rate of the public to growth rate of private funding – if this ratio is more than one, it has a positive effect on the financial flexibility of free economic zones and characterises their ability to change;

3.2. Indicators of reaction to the price changes show the degree of elasticity of demand in relation to changes in the indicative cost of tuition, cost of fee-paying students and the average tuition fee of a contract student. Suppose the university meets the crisis with a high tuition fee. In that case, it negatively affects the ability to manoeuvre prices, the formation of a new price to attract contract students during the financial crisis caused by Covid-19, reduces the competitiveness of high education institutions. Such indicators include:

- the ratio of the average cost of fee-paying students to the average cost of public funding education of the student
- the ratio of indicative cost to the cost of education of a public funding student
- the ratio of the average tuition fee of a contract student to the average cost of education of a public funding student

4. Cost-effectiveness indicators show whether universities are able to adapt to the challenges of the Covid-19 pandemic, which is manifested in the positive values of indicators and their dynamics:

- increase in surplus / deficit - a positive value and increase in the dynamics is a stimulating indicator of financial flexibility of the university;
- net cash flow ratio is calculated as the ratio of net cash flow to the assets, shows the degree of replenishment of funds due to net cash flow;
- the increase in equity and financial results is the result of practical activities of the university, which is ensured by balancing revenues and expenditures, including optimisation in the period of Covid-19.

The assessment of the level of financial flexibility according to the system of proposed indicators should be calculated based on an integrated indicator by standardising indicators and ranking them according to the degree of impact. The conditions were met during rationing: for the stimulating indicators, the increase of which leads to an increase in financial flexibility, the minimum value was applied; for the destimulators the increase which leads to a decrease in flexibility, the maximum value is applied.

For the evaluation of financial flexibility, the official data of 24 universities of Ukraine (classical (CU), economic (EU), polytechnic (PU), pedagogical (PedU), others (OU)) and 19 indicators were applied. Indicators for the calculation of which are missing in the official financial statements and are internal management information (administrative costs to total income, the utility costs to fixed costs and indicators of response to the price factor) in the calculation of integrated indicator were omitted. As a scale indicator, the contingent of universities on the first of October of the respective year was taken [7]. When determining the integrated indicator, its logarithm was performed, and the decimal logarithm was calculated to bring the data to a single array.

The integrated indicator was determined separately for 2019 and 2020; the average value of the integrated indicator and rationing was carried out over a set of years. The results of the integrated indicator are presented in Figure 2.

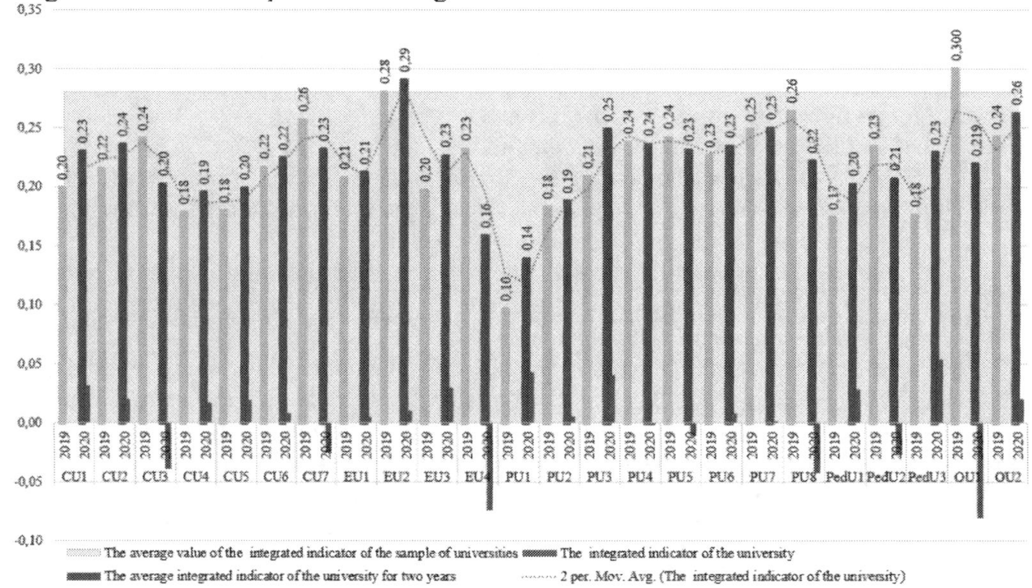

Fig. 2. An integrated indicator of the financial flexibility of universities

The obtained average value of the integrated indicator for 2019-2020 is 0.28, so we will assume that universities with a value above the average are more financially flexible than others. The highest level of financial flexibility among classical universities in 2019 is demonstrated by CU7, which is 0.40. However, in 2020 the integrated indicator decreased by 0.06 due to a reduction in the scale (reducing the number of students by 22%), which negatively affected growth rates of public and private funding, increasing the budget deficit and the share of staff costs to total costs. The level of financial flexibility of CU2 increases during 2019-2020 from 0.27 to 0.29 due to a positive increase in the contingent by 2%, the transition to a surplus zone, a positive increase in equity and financial results, an increase in the replenishment ratio by eight times. The high level of the integrated indicator for CU3 was set at 0.30 in 2019, but the effects of Covid-19 violated the level of financial flexibility by 0.05 to 0.25.9 The increase in the scale by 6% with a simultaneous change in the structure by 3% towards the growth of private funding by 0.02, affected the growth rate of private funding by 0.5%, reducing the growth rate of budget funding by 16.6%. An insufficient level of financial flexibility in the group of classical universities is demonstrated by CU1, CU4, CU5, the level of the integrated indicator, despite the positive dynamics, is below the average value and is 0.23 in 2020; 0.25; 0.27 respectively. This is due to the reduction of scale indicators, growth rates of public financing, growth of equity and financial results, the budget deficit, fluctuations in other indicators.

The financial flexibility of economic universities depends on the indicators of the reaction to the price factor because the gap between the cost of education of public funding students, a fee-pay student and the indicative cost significantly affect universities' income. The change in scale ambiguously affects the integrated indicator, and therefore for economic universities, the scale indicator is not a determining factor, but to a greater extent, is determined by its structure and cost of education. For example, despite an increase in scale by 11%, the financial flexibility of the EU4 decreased from 0.30 to 0.22 due to a reduction in the growth rate of private funding and income of the university. The reduction of the contingent by 5% in the EU3 did not have a negative impact on the financial flexibility of the university, which increased during 2019-2020 from 0.24 to 0.28, as the reduction of private funding growth while increasing the public funding had a positive impact on the integrated indicator, that confirms the influence of price indicators.

The high level of financial flexibility is inherent in polytechnic universities, PU1-PU7 show a positive increase in the integrated indicator, but PU1 and PU2 have an insufficient level of financial flexibility, which is in 2020 are 0.19 and 0.25, respectively. This is due to a slight reduction in the contingent, a decrease in the growth rate of private and budget funding. The high share of public funding, scale and cost of the public funding student are positive factors of financial flexibility.

The integrated indicator of financial flexibility for pedagogical universities is below the average of 0.26-0.25 in 2020. For the other group, it includes the National University of Food Technologies (IU1) and the National University of Life and Environmental Sciences of Ukraine (IU2), which demonstrate above the average of the sample of universities, the level of the integrated indicator in 2020 is 0.29 and 0.30, respectively.

Assessment of financial flexibility indicators for the sample of universities indicates an insufficient level of financial management in the high education institutions, which with the expansion of autonomy do not carry out medium-term planning of financial resources, evidence of sharp fluctuations, budget deficit, negative net cash flow, etc. The most important indicator influencing the financial flexibility and sustainability of universities is the contingent, which during the pandemic period was significantly reduced in regional institutions. Given that the financing of Ukrainian high education institutions is tied to the contingent, the last one has significantly affected their financial support. These trends disrupt the financial balance of the universities and reduce the level of financial flexibility during the period of adaptation to Covid-19.

The results of the correlation analysis showed that the degree of connection between the indicators of financial flexibility and the integrated indicator is not high and not more than 0.6 (Fig. 3).

A significant correlation at the level of 0.40-0.52 is observed with respect to autonomy ratios, the ratio of net assets to assets, liquidity ratios. A weak negative correlation between

-0.10 and -0.27 is observed between the integrated indicator and the marginal revenue indicator, the share of staff costs to total costs and the share of private financing. For other indicators, there is a weak correlation between 0.38 and 0.38. The results of the correlation analysis showed significant numbers of correlations of the same indicators in 2019-2020, which is a sign of the sensitivity of universities to the effects of Covid-19.

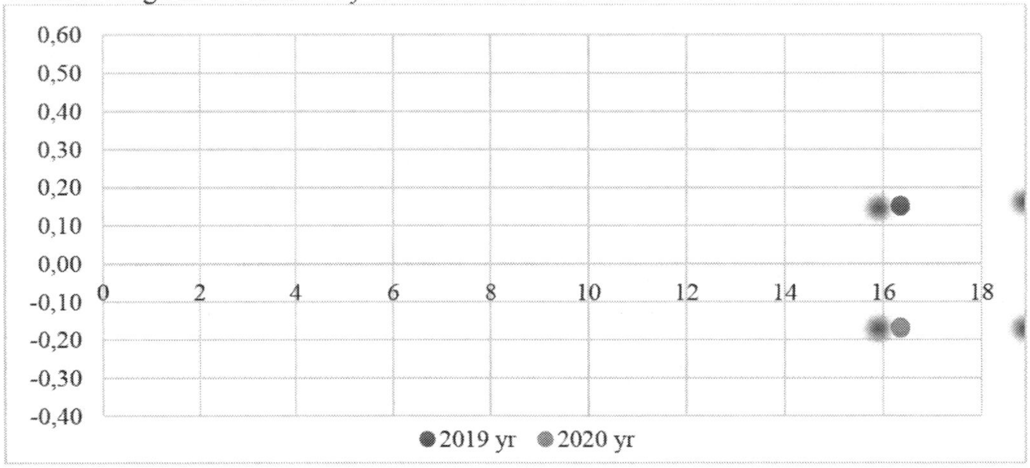

Fig. 3. Results of correlation analysis of financial flexibility indicators

Thus, the financial flexibility of the university is determined not only by quantitative indicators but also qualitative, the ability of university management to make decisions about the formation of the required value proposals in terms of adaptation to Covid-19. Decision-making is carried out under the influence of external and internal factors, which in turn are divided into productive (direct, indirect) and behavioural [8]. Behavioural factors significantly affect the level of financial flexibility, namely the effectiveness of university management, the ability to interact with stakeholders, academic reputation, quality of organisation during the adaptation to Covid-19, irrationality of expectations, motivation of potential applicants, regional factor, effect scale, location of the university, etc. Regional universities are more sensitive to the crisis than central ones.

Conclusions

The financial flexibility of universities in adapting to Covid-19 has different manifestations in terms of public and private funding. Ukrainian universities are financed mainly by public allocations, which depend on the number of students and performance indicators, which reduces the level of financial flexibility, which is manifested through the ability of educational institutions to restrain and optimally allocate costs. Regarding private funding, their share in universities ranges from 17 to 72%, with an increase in the share of economic universities, which poses particular challenges to their resources. To maintain a stable position and ensure expanded reproduction, it is necessary to maintain leading positions in the market of educational services, provide quality educational services.

The results of the practical approbation showed that the clear indicators for the financial flexibility of universities are changes in the volume and structure of the contingent, not only in terms of public and private funding but also in terms of specialities, taking into account the price factor (tuition fees and contract prices). The results of the analysis revealed that the reduction of the contingent and the change in the structure of funding in different ways affect the change of the integrated indicator of financial flexibility for different groups of universities due to the list of specialities and their cost.

The level of financial flexibility of universities and its manifestation is the ability to generate their own income, generate value proposals, interact with the business environment, respond quickly and adapt to crisis conditions, implement business models, act in new situations.

The proposed methodological tools for assessing the financial flexibility of the university in terms of adaptation to Covid-19 considers the signs of financial flexibility, distinguish four groups of indicators with indicators for their evaluation: sensitivity, efficiency, manoeuvrability, economy.

The methodological tools were tested in a sample of 24 universities, financial flexibility indicators were calculated, and an integrated financial flexibility indicator at the Covid-19 input was determined, taking into account the consequences of the impact. It was found that the ambiguous dynamics and importance of indicators of financial flexibility, manifestations of atypical for some universities financial indicators are due to changes in the structure of funding and the contingent of students. Such trends have affected financial flexibility and so on the financial results of universities in different ways.

Further research should be aimed at modelling the effects of Covid-19, considering the price indicator and changing the structure of the contingent on the financial stability of universities as an indicator of financial flexibility.

References

1. O. Laktionova. *Flexibility of the financial system: methodology, evaluation and vectors of provision.* (Vinnytsia, Ukraine, 2016).
2. J. Ang, A. Smedema, Journal of corporate Finance, **17**(3), (774-787). (2011). DOI: 10.1016/j.jcorpfin.2011.02.001.
3. A. Gamba, A. Triantis, Forthcoming Journal of Finance, **63** (5), (2008). DOI: 10.1111/j.1540-6261.2008.01397.x.
4. Yaw M. Mensah, R. Werner, Journal of Accounting and Public Policy **22** (4), (293-323). (2003). DOI: 10.1016/S0278-4254(03)00036-X.
5. R. Islam, M. Wang, L. Dewri, International Journal of Accounting and Financial Reporting, **9**, (245-256). (2019). DOI: 10.5296/ijafr.v9i1.13195.
6. R. Fahlenbrach, K. Rageth, R. M Stulz, The Review of Financial Studies, **34** (11), (5474-5521). (2020). DOI: 10.1093/rfs/hhaa134.
7. *The only state electronic database on education*, 2021. URL: https://registry.edbo.gov.ua/opendata/educators/
8. L Shaulska, L. Yurchyshena, Y. Popovskyi. *Using MS Power BI tools in the university management system to deepen the value proposition,* in Procceding of the 11th International Conference on Advanced Computer Information Technologies, 2021, ACIT (2021). DOI: 10.1109/ACIT52158.2021.9548447.

Changes in the Structure of International Trade And Services in the Context of Tax Fraud and Tax Evasion

Antonín Korauš[1]*, *Miroslav* Gombár[2], *and Filip* Černák[2]

[1]Academy of the Police Force in Bratislava, Sklabinská 1, 835 17 Bratislava 35, Slovak Republic
[2]University of Prešov in Prešov, Faculty of Management, Konštantínova 16, 080 01 Prešov, Slovak Republic

Abstract. The EU and the Member States have been forced in recent years to take a stronger position against the growing trend of tax fraud, tax evasion, and tax avoidance. The EU legislative and non-legislative acts in this area have resulted from political attitudes, economic intentions, proposals, and compromises reached between the individual Member States, on the one hand, and the EU, on the other, in their effort to combat fraudulent behaviour in the taxation area. The fight against fraudulent behaviour in the taxation area, be it tax fraud, tax evasion or tax avoidance, has become a real global challenge not only for the EU and its Member States but also for the entire world. These are different forms and methods for misusing tax systems both within the EU and abroad. This paper is part of the project: "Electronic methods for detecting unusual business transactions in the business environment" (ITMS code: 313022W057), co-financed by the European Regional Development Fund (ERDF) and also the project "VEGA1 / 0194/19 — Research on process-oriented financial management with a focus on to detect tax evasion in international trade".
Keywords: international trade, international services, tax fraud, tax evasion.

1 Introduction

The fight against the misuse of the tax system in the eyes of the EU law is undoubtedly a very up-to-date topic and, at the same time, a very complex and sensitive one. This applies to indirect taxes (in particular VAT) and direct taxation [1, 2]. In this context, it should be emphasised that ever since the European Commission has defined the terms tax fraud and tax evasion, the notion of tax fraud seems to be "linked" to the VAT system, and also consumer taxes, or more aptly, it is linked to the misuse of the indirect taxation system. Consequently, tax evasion is "directly linked" to the income tax [3].

In the context of VAT tax fraud, this is ultimately a evasion of VAT tax, thus causing losses in the state budget revenue. Such evasion, however, may not necessarily result from

* Corresponding author: antonin.koraus@akademiapz.sk

the tax fraud committed by a tax entity within the applicable system of VAT application, which is the object of interest for the criminal law.

A similar situation applies when considering tax evasions resulting from a tax entity's illegal conduct or behaviour. In this case, too, we may state that, in the eyes of the tax law and even of the criminal law, this is a tax fraud de facto, even though it is not directly related to the issue of indirect taxation. It is pretty understandable that the term tax fraud applies to the VAT system. As pointed out by the European Commission, the EU VAT system benefits the single market, as it has removed barriers distorting competition and hindering the free movement of goods, thus greatly facilitating trade in the EU [4, 19]. As a broad-based consumption tax, VAT is considered one of the best forms of taxation that stimulates growth. It is also a significant and growing source of tax revenue within the EU [5].

2 Management of the elimination of severe tax evasions and fraud

Given that the management of the elimination of severe tax evasions and fraud is linked to the European territory and concerning the importance of this issue in the European and global context, it is necessary to look for the causes and existence of tax evasions and then look for ways to eliminate them using appropriate management measures [6, 18]. The importance within the society of the theoretical and subsequent practical outputs in this area has been growing and affects the economics, law, economy, and international rating of countries.

Whatever aspect we chose to assess the quality and effectiveness of the fight against misuse of the tax system, we would always come to the following conclusions:

- misuse of the tax system consists of separately perceived infringements that take tax fraud, tax evasion, and tax avoidance. In most cases, the nature of these infringements is in contrast with the rules laid down in the relevant tax legislation which govern the prohibition of the misuse of tax law;
- in practical terms, tax fraud may not be separated from tax evasion or tax avoidance. In this context, it should be emphasised that, in the eyes of the criminal law, tax fraud is understood as intentional and illegal tax evasion;
- due mainly to the political decisions taken by the European Commission, tax frauds are linked to the issue of indirect taxation, and they also affect the amount of state budget revenue from taxes in any state due to income tax;
- VAT tax frauds are not/do not have to necessarily result from the tax fraud committed by a tax entity. In this context, it can be stated that even an illegal tax evasion is, a de facto tax fraud in the eyes of the tax law.
- in an economic and political alliance, such as the EU, one cannot view the concerted EU efforts to combat VAT fraud separately from the Member States' efforts in this area. Without clearly demonstrated will of all EU Member States, the Union as a whole would not have achieved any "more tangible" success in the fight against tax fraud, tax evasion and other tax infringements;
- the interests of EU Member States in combatting the misuse of the tax system often differ from the EU interests, as demonstrated by the complex nature of the procedures to reach unanimous agreement when adopting EU legislation in this area;
- it would be appropriate to explicitly state in the EU legislative and non-legislative acts, in the political documents issued by the public authorities, in explanatory memoranda to tax laws, etc., that a distinction must be made for the term "tax evasion":

▪ tax evasion sensu stricto, as a result of illegal/legal conduct and behaviour of tax entities in the field of income taxation, by which the tax entities reduce/eliminate their tax liability to the state budget;

▪ tax evasion sensu largo, as a result of reduction/loss of state budget tax revenues due to tax fraud, tax evasion sensu stricto and/or tax avoidance.

Tax fraud has posed a massive problem in the operation of the EU throughout the EU's existence which continues to this day. This particularly applies to VAT frauds which undermine the smooth operation of the EU internal market [7]. On the other hand, it would be very misleading for us to conclude that the EU institutions have neglected their responsibilities in the fight against tax fraud. Instead, their interest was not primarily focused on taking adequate measures to prevent tax fraud, and they focused on the subsequent resolution of the ensuing problems [8, 20].

Regardless of the subject of the taxation, VAT tax fraud may take various forms and features. These may include wilful conduct or inaction of the tax entity (e.g. by their failure to file a tax return, etc.). However, VAT frauds are primarily associated with tax entities' active and mostly intentional conduct [9, 17].

According to a report published by the European Court of Auditors (ECA), the exchange of tax information between the Member States is still insufficient to ensure fair and efficient taxation within the entire single market. There are problems not only in the EU legislative framework but also in its implementation and monitoring. In particular, the auditors found numerous cases of poor quality of the information being exchanged or that the information was not utilised to the necessary extent.

The ever-increasing number of cross-border transactions complicates the proper evaluation of tax obligations by the Member States and provides room for avoiding tax obligations and for tax evasion. The revenue losses from tax avoidance alone by legal entities are estimated to range from € 50 billion to € 70 billion in the EU. If special tax rules and inefficient tax collection are included in this estimate, the above amount will increase to almost € 190 billion. Therefore, cooperation between the Member States is essential to collect the total amount of taxes on time.

Value-added tax is one of the indirect taxes, which means that this tax is paid indirectly, i.e. in the price of goods and services. The specific nature of the value-added tax lies in the fact that the entity which collects and pays the tax to the state budget (i.e. the taxpayer) and the entity which pays this tax is not the same person, namely because:

• the entity collecting and paying this tax is a legal or natural person - entrepreneur (VAT payer),

• the entity paying this tax is the final consumer (customer) who pays the VAT in the price of the goods or services.

Every year in the EU, billions of euros are lost to tax evasion. Efforts by national tax authorities to tackle this are increasingly hampered by new business models, especially those based on digital technology.

Tax-compliant businesses also need a more straightforward and more up-to-date tax system if they are to benefit from the single market. This action plan presented key initiatives to:

• tackle tax fraud,

• make compliance easier,

• take advantage of the latest developments in technology and digitalisation.

According to a new special report published by the European Court of Auditors (ECA), there is still insufficient sharing of tax information between the EU Member States to ensure fair and effective taxation throughout the Single Market. The problems are not only with the EU's legislative framework but also with its implementation and monitoring. In particular, the auditors found that, often, the information exchanged is of limited quality or underused. The ever-growing number of cross-border transactions makes it difficult for the Member States to assess taxes due properly and encourages tax avoidance and evasion. Revenues lost to corporate tax avoidance alone are estimated at between €50 billion and €70 billion yearly in the EU, reaching some €190 billion if special tax arrangements and tax collection inefficiencies are included. Cooperation between the Member States is essential to make sure taxes are collected in full and due.

The legislative framework the European Commission has established for the exchange of tax information is transparent and logical. However, it suffers from several gaps, warn the auditors. Firstly, it remains incomplete concerning stemming tax avoidance and evasion. Cryptocurrencies, but also other forms of income, for instance, are not subject to mandatory reporting, thus remaining largely untaxed. Secondly, the support provided to the Member States does not go far enough. In particular, the Commission barely addresses the issue of poor data quality and does not assess how effective and deterrent the sanctions for non-compliance are. Finally, the Commission should provide more guidance to help the Member States, especially in data analysis and use [10].

The national governments of the EU Member States are broadly free to design their tax laws and systems, and decision making in this policy sensitive area at an EU level requires the agreement of all Member States. However, all national legislation must respect certain fundamental principles, such as non-discrimination and freedom of movement in the internal market. The tax laws of one country should not allow people to escape taxation in another.

Member States in their turn must:

- operate the taxation systems;
- collect and report the required tax-related information;
- share best practices;
- ensure fair tax competition in the internal market.

Many countries have agreed to lift bank secrecy for tax purposes, share tax information with other governments and remove the secrecy surrounding the owners, banking arrangements and financial transactions of legal entities. The starting point was the global EOIR standard developed by the OECD, which first provided for information to be exchanged internationally between two countries at the tax authority's request, with a new requirement added to obtain and exchange information even if held by a bank. Technological advances have also led to the widespread automatic exchange of information, complementing EOIR.

3 Conclusion

The implemented legislative measures also form an essential aspect of the elimination of tax evasion in that they helped transform the behaviour of taxpayers, increase their discipline and help form their responsible approach to the selection of business partners. As a negative side effect, the approaches and variability of tax evasion and frauds have become more diverse and more sophisticated.

Legislative tax measures have translated into an increase in state budget revenues, which positively affected the achievement of fiscal policy objectives, especially in reducing the GDP deficit. Increasing the efficiency of tax collection and the identification and elimination of tax evasion and fraud is a long-term process. The importance of adopting a permanent and effective solution to address tax evasion and fraud is possible through a combination of cooperation and legislative measures with an emphasis on tax law. Legislative tax standards create room for implementing tax planning, tax optimisation and, from an opposing point of view, this also provides an opportunity for aggressive tax planning, even creation and use of schemes to commit tax evasion and fraud[11].

Over the past years, the EU and the Member States have been forced to take a stronger position against the growing trend of tax fraud, tax evasion and tax avoidance. [12] The EU legislative and non-legislative acts in this area have resulted from political attitudes, economic intentions, proposals and compromises reached between the individual Member States, on the one hand, and the EU, on the other, in their effort to combat tax fraud [13].

As stated by [14], the fight against tax fraud, be it tax fraud, tax evasion or tax avoidance, has become a real global challenge not only for the EU and its Member States but also for the entire world. These are different forms and methods for misusing tax systems both within the EU and abroad.

In the eyes of the EU law, the fight against the misuse of the tax system is currently a crucial, complex and susceptible topic. This does not apply to indirect taxes only, in particular VAT, but also to direct taxation.

Tax fraud and tax evasion have been and are likely to continue to represent a significant issue throughout the existence of the EU [15]. This is particularly true in VAT, as this issue distorts the law's correct and well-functioning EU internal market. However, we cannot state that the EU institutions have neglected their obligations to fight against tax fraud. Rather than addressing and preventing causes, the institutions dealt with the consequences. Adequate measures were not taken as a matter of priority to prevent tax fraud; instead, they focused on the consequences and thus addressed the following issues. It should be emphasised that VAT evasion is a fundamental problem in tax administration regarding the scope, forms, methods or procedures and one that needs to be resolved by the tax administrations in the individual EU Member States.

At the beginning of 2020, the world was hit by a situation which no one expected. Coronavirus pandemic has affected most countries around the globe and caused significant changes in the economy and social life of the countries and in the life of every single citizen. The pandemic seriously disrupted and changed conditions for doing business, the structure and volume of services worldwide. E-commerce is no exception, even though at first glance, it might seem that it was and will be affected to a lesser extent than other areas. The operations of numerous companies in China, Europe or the US have virtually stopped, even though e-commerce continues to operate and the pandemic has had a different impact on its functioning [16]. At the same time, various general measures are adopted in individual countries, which follow up on their risk management strategies, critical security assets, risk mitigation priorities, etc. Subsequently, the strategies are re-evaluated in the area of healthcare, providing supplies to the general public, conditions of doing business, mobility and provision of services.

Based on the available information, we can see that the coronavirus pandemic affects the shopping behaviour of customers and corporate entities. The search rankings have changed significantly for the most popular products in the e-commerce of natural persons. The focus of attention has moved to products that are most important for people's lives (food) or those helping to overcome quarantine or social exclusion and restrictions on freedom of movement (board games, sports equipment, etc.).

In general, there has been an increase in the online sales of products, although it is not yet possible to recognise a clear trend in the long run. However, we can state that the customers started to use online shopping to a greater extent. From a consumer behaviour perspective, this may also result in increased trust in online shopping, potentially leading to e-commerce revenue growth even after the coronavirus pandemic. However, to analyse every impact of the pandemic on e-commerce, we will need to explore additional changes in the economy and social life of individual countries and put them in relation to the time, space and content of individual quarantine measures, especially as regards the security of activities in the online environment. In the coronavirus pandemic, there is an increased possibility for the misuse of cyberspace for unfair and fraudulent practices and activities, which the perpetrators of such activities also used. The business environment is no exception, and the number of cyberattacks, fraud and other criminal activities is on the rise.

This publication has been written thanks to the support of the Operational Programme Integrated Infrastructure for the project: "Electronic methods for detecting unusual business transactions in a business environment" (ITMS code: 313022W057), co-funded by the European Regional Development Fund (ERDF) and also by the "VEGA1/0194/19—Research on process-oriented management of financial management focusing on detection of tax evasion in terms of international business".

References

1. B. Salanié, The Economic of Taxation, (MIT Press, Cambridge, UK, ISBN: 0-262-19486-4, 2003).
2. M. Schratzenstaller, Obs. Diagn. Econ. **141**, 55–77. https://www.ofce.sciences-po.fr/pdf/revue/3-141.pdf. (2015)
3. K. Krzikallová, F. Tošenovský, Sustainability, **12**, 4925. https://doi.org/10.3390/su12124925 . (2020)
4. *European Commission :VAT Rates Applied in the Member States of the European Union.* Situation at 1st January 2020; EuropeanCommission: Brussels, Belgium. (2020)
5. M.Keen, B. Lockwood, Journal of Development Economics. **92**(2), 138-151.DOI: 10.1016/j.jdeveco.2009.01.012 (2009)
6. S. Cnossen, International Tax and Public Finance, **25**, 519–551. https://doi.org/10.1007/s10797-017-9453-4 (2018)
7. M. Bonucchi, M.Ferrari, S.Tomasini, T.Tsvetomira, Revue de l'OFCE, **141**(5), 225–262. https://www.cairn.info/revue-de-l-ofce-2015-5-page-225.htm#citepar (2015)
8. A. Kowal, Zeszyty Naukowe Małopolskiej Wyższej Szkoły Ekonomicznej W Tarnowie, **44**(4), 39-51. https://doi.org/10.25944/znmwse.2019.04.3951 (2019).
9. G. Sobiecki, M. Szwed-Ziemichód, Analizy i Studia **1**, 3–33. DOI: https://doi.org/10.33119/ASCASP.2019.1.1 (2019)
10. Special report 03/2021: "Exchanging tax information in the EU: solid foundation, cracks in implementation" is available on the ECA website (eca.europa.eu) in 23 EU languages. https://op.europa.eu/webpub/eca/special-reports/tax-03-2021/sk/ (2021)
11. A. Andrejovská, V. Konečná, J. Hakalová, Ad Alta J. Interdiscip. Res. **10**, 8–13. ISSN: 1804-7890, 2464-6733. (2020).
12. M. Feldek, Financial Law Review, **20**, 35-52. ISSN: 2299-6834. DOI: 10.4467/22996834FLR.20.019.13091 . (2020).
13. G. Poniatowski, M.Bonch-Osmolovskiy, J.M.Durán-Cabré, A.Esteller-Moré, A. Smietanka, Studyand Reports on the VAT Gap in the EU-28 Member States: Final Report. Available online: http://case-research.eu/files//?id_plik=5692. (2018).
14. M.Bernasconi, L.Corazzini, R. Seri, Journal of Economic Psychology, **40**, 103–118. DOI: 10.1016/j.joep.2013.01.005 . (2014).
15. B. Timmermans, W.M.J. Achten. Int J Life Cycle Assess **23**, 2217–2247. https://doi.org/10.1007/s11367-018-1439-7. (2018).
16. M.Sikandar. Mapping The Impact Of COVID-19 Pandemic On E-Commerce. Available at: https://cedcommerce.com/blog/impact-of-covid-19-on-ecommerce/ (2020).

17. L. Mura, J. Buleca, Procedia Economics and Finance, **3**, pp.217-222. (2012)
18. L. Mura, Z. Hajduová, Entrepreneurship and Sustainability Issues, **8(3)**, 487-503. https://doi.org/10.9770/jesi.2021.8.3(31). (2021).
19. M. Srebalova, M. Horvath, J. Vacok, F. Vojtech, S. Filip, Entrepreneurship and Sustainability Issues, **7(4)**, 3385-3394, https://doi.org/10.9770/jesi.2020.7.4(53). (2020).
20. D. Novackova, T. Peracek, L. Strazovska, B. Much, Juridical Tribune - Tribuna Juridica, **10**, 142-163. (2020).
21. O. Prokopenko et al. International Journal of Scientific and Technology Research, **9(3)**, pp. 1112-1115. (2020).
22. O. Prokopenko, V. Omelyanenko, TEM Journal, **9 (3)**, pp. 1068-1075. DOI: 10.18421/TEM93-31. (2020).

AUTHOR INDEX